The Orchids

Harvard University Press

CAMBRIDGE, MASSACHUSETTS, AND

LONDON, ENGLAND   1981

The Orchids

NATURAL HISTORY AND CLASSIFICATION

ROBERT L. DRESSLER

Copyright © 1981 by the Smithsonian Institution
All rights reserved
Printed in the United States of America

Library of Congress Cataloging in Publication Data
Dressler, Robert L   1927–
    The orchids.
    Includes bibliographical references and index.
    1. Orchids.   2. Orchids—Classification.
3. Botany—Classification.   I. Title.
QK495.064D69   584'.15   80–24561
ISBN 0–674–87525–7

climates for which they were written. Neither will this book tell how to identify orchids to species, though the need for such books is quite clear.

Although the idea of some sort of a botany of the Orchidaceae had been in my mind for years, the resolution to write a book of this scope and content was conceived on the evening of July 8, 1978, while I was chatting with Dr. Norris H. Williams, and actual writing started a few hours after I left him at the airport the next day. Like the creations of Agatha Christie's Ariadne Oliver, the book quickly acquired a life and singularly dictatorial character of its own. It would not permit work on, or even thought about, other projects for more than a few minutes at a time. Even field trips had to be put off until the first draft was finished. Under its influence, orchid facts which had been lying about in plain sight for years suddenly became aggressive and demanded attention and explanation.

This book is, in part, descended from a short course on orchids given at the Museo Departamental de Historia Natural in Cali, Colombia, in December 1975, in collaboration with Dr. Alvaro Arango, and a similar course given in Panama in October 1977 under the auspices of the Museo de Ciencias Naturales. There is also a bit of genetic material from a short course in floral ecology given at the Universidad Nacional Autónoma de México in May 1970, and a more tenuous relationship with a course in biogeography given at Washington University in St. Louis, Missouri. To the sponsors of these courses and to all the students who asked questions, my sincere gratitude. I presented some of the ideas which have developed into the classification used here at the Twelfth International Botanical Congress in Leningrad.

I have tried to incorporate into my discussions material from papers that have come to my hand before October 1979. I am sure that I have overlooked some important publications, through either ignorance or inefficiency. I am gratified (and a bit dismayed) by the amount of new data that has come to hand while this book was in press. The latest (unpublished) information on seed structure from Barthlott and Ziegler, for example, suggests that (1) the Dendrobiinae are not very closely related to the Eriinae, and the Dendrobiinae, Bulbophyllinae, Coelogyninae, Glomerinae, and Thuniinae may form a very natural group (tribe Dendrobieae?), and (2) the vandoid orchids are not really a natural phyletic group, but have evolved from several different epidendroid groups. Clearly, more study is needed.

I am indebted to many people for help during the writing of this book. First and foremost, I must thank my wife, Kerry, for her tolerance and encouragement, as well as for her help as photographer and field companion. James D. Ackerman, C. H. Dodson, K. S. Walter, C. A. Luer, P. Taylor, and N. H. Williams each read a preliminary draft of the book, and all made valuable suggestions. Others with whom I have discussed some of the material presented here include J. Atwood, W. Barthlott, P. J. Cribb, J. P. Folsom, L. A. Garay, P. S. Lavarack, G. Seidenfaden, K. Senghas, R. Silberglied, and L. Y. Th. Westra. For the color plates I have tried to select photographs that show something of the diversity of the Orchidaceae. A number of friends and colleagues have been most generous in lending slides for this purpose, and I am indebted to W. Barthlott (figures 40, 50), C. H. Dodson (20, 92), R. Escobar R. (95), J. P. Folsom (19), J. Green (16, 35, 45), M. W. Hodge (72), R. Jenny (3, 70), H. A. Kennedy (13, 59, 66, 89), J. Kuhn (86), P. Lavarack (8, 12, 29, 34, 44), C. A. Luer (2, 4, 7, 9, 25, 30, 31, 32, 42, 49, 51), E. A. Schelpe (41, 82), F. L. Stevenson (18), J. Stewart (36, 39, 78), W. P. Stoutamire (21, 22, 23), K. W. Tan (14), P. Tonelly (33), and K. W. Walter (17, 24, 26, 61, 71, 75, 84, 85). Arlee Montalvo's skill and patience in the preparation of drawings are deeply appreciated. Many others have helped me to find orchids in the field or have given freely of materials from their own gardens. A complete list of these contributors would be much too long and tedious, but I am deeply indebted to all of them.

R.L.D.

Contents

The Orchids

What Orchids Are

1

The Orchidaceae is one of the largest and most diverse families of plants, including somewhere between one tenth and one fourteenth of all flowering plant species. In this actively evolving group, highly specialized adaptations for attracting, deceiving, and manipulating insects to achieve cross-pollination have fascinated observers since the time of Darwin. Hobbyists especially have been attracted to orchids, cultivating them in greenhouses, gardens, windows, and basements around the world. One might expect the orchids to be among the most scientifically studied and best understood of plants, but that is far from the case. Many botanists seem to feel that orchids are unsuitable objects for professional study, or else that everything must have been said already about such a popular plant group. How wrong this is becomes quickly evident to anyone who tries to collect and identify orchids in the tropics.

We may begin to understand what orchids are by first considering their place in the world of living things. In years past all terrestrial life was customarily divided into the plant kingdom and the animal kingdom. But because some living things, particularly one-celled organisms, do not fit neatly into one category or the other, it has been suggested recently that four or five "kingdoms" would give a clearer idea of the diversity of terrestrial life (see fig. 1.1). As long as we keep in mind that these kingdoms are concepts of convenience rather than sharply defined phyletic units, such a five-kingdom system is very useful. If, on the other hand, we wanted sharply defined phyletic units, then the three lobes of the diagram in figure 1.1 would become at least eighteen different kingdoms, and plants, fungi, and animals would become—again for the sake of convenience—"superkingdoms." Viewing just the plants in this way, we can find three main groups: the red algae, the brown algae and their allies, and the green plants, which include a vast array, from simple, one-

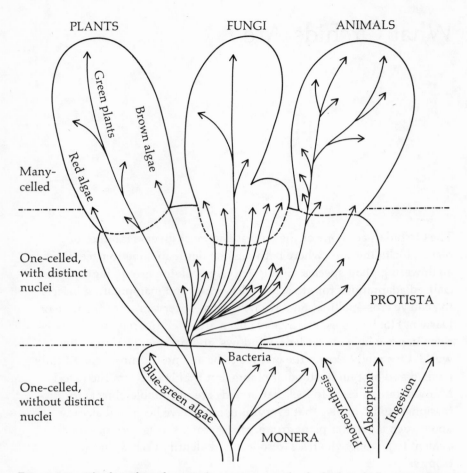

Figure 1.1 The five "kingdoms" of organisms, each of which includes several or many phyletic lines. (After Whittaker, 1969.)

celled organisms up to the giant redwoods (fig. 1.2). Our interest here is in the flowering plants, which make up the majority of the plants we see, grow, and use in day-to-day life.

The class Angiospermae, or the flowering plants, is customarily divided into two subclasses, the Monocotyledonae (monocots) and the Dicotyledonae (dicots). The major differences between these two groups are shown in table 1.1. Some plants, however, will not show all of these features. The Araceae usually have net-veined leaves, and most orchids do not have any cotyledons, yet both are considered monocots. While most botanists treat these two subclasses as "real" groups, Huber (1977) suggests that they are, rather, two poorly defined segments of a single,

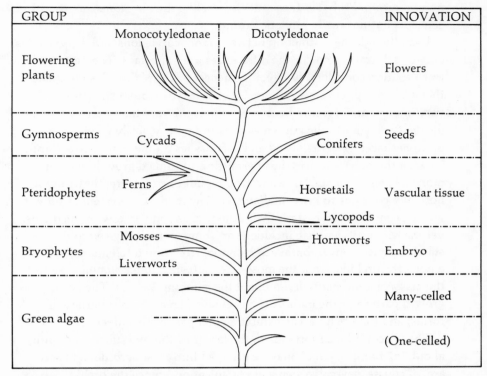

GROUP		INNOVATION

Figure 1.2 A diagrammatic family tree of the green plants, showing levels of evolution. The "innovations" shown on the right are the main evolutionary developments that characterize each level (also present in subsequent levels). The flowering plants, or angiosperms, represent a natural phyletic group, but the bryophytes, pteridophytes, and gymnosperms represent levels of evolution, rather than natural groups.

Table 1.1. A comparison of the Monocotyledonae and the Dicotyledonae.

Monocots	Dicots
Cotyledons (embryonic seed leaves) 1	Cotyledons usually 2
Leaves mostly parallel-veined	Leaves mostly net-veined
Vascular cambium lacking	Vascular cambium usually present
Vascular bundles generally scattered, or in two or more rings	Vascular bundles usually in a ring which encloses pith
Flower parts, if definite in number, in 3s, rarely in 4s or 5s	Flower parts, if definite in number, in 5s, sometimes in 4s, seldom in 3s
Pollen grains typically with a single furrow or pore	Pollen grains typically with 3 furrows or pores

natural group. Probably these subclasses, like the kingdoms, are useful concepts but not really phyletic groups.

Even though the distinction between monocotyledons and dicotyledons may be arbitrary, some of the differences are important. The patterns of leaf venation considerably affect the aspect of the plants, despite the fact that each group has a great diversity of leaf types. Stem structure is a very important difference because it affects the way that plants grow: in dicots, the layer of cells known as the cambium typically produces wood or, in herbaceous stems, wood and soft tissues on the inner surface and bark on the outer surface, thus permitting continued growth in diameter; monocot stems may be large and woody, but once mature, they cannot normally continue to grow in diameter. The monocots' system of growth would seem to severely limit their potential as compared with the dicots, yet the Monocotyledonae include everything from duckweed to the stately palms, from crabgrass to bamboo, from onion to banana, and, among the orchids, from the moss-like *Platystele jungermannioides* to the massive *Grammatophyllum* and the climbing *Vanilla*. The basic growth pattern of the monocots has evolved an enormous variety of forms, and the Orchidaceae exhibit a good part of this diversity.

The Dicotyledonae form much the larger of the two subclasses, with about 267 families which may be grouped into 19 superorders. (There are, of course, several systems of classification.) Since the orchids fall into the Monocotyledonae, we need not go into greater detail about the dicots. The Monocotyledonae are usually divided into four or five super-orders, though there is as yet little agreement either on the delineation of the superorders or on their nomenclature; for example, different authors use the names Liliiflorae, Liliidae, or Lilianae for the superorder to which the orchids belong. I will use the system of superorders outlined here (see fig. 1.3):

(1) Alismatiflorae: mostly aquatic or semiaquatic herbs with many stamens and carpels.

(2) Ariflorae: usually herbaceous or climbing plants with net-veined leaves and small, densely clustered flowers. The Araceae and Lemnaceae are in this superorder.

(3) Areciflorae: a major group with large, palm-like leaves and individual flowers that are usually small and insignificant. This superorder includes the palms and the Cyclanthaceae.

(4) Commeliniflorae: another major group in which the outer perianth parts (sepals) are green and leaf-like, and the inner ones (petals), when present, are quite distinct. The bromeliads, gingers, bananas, and grasses belong here.

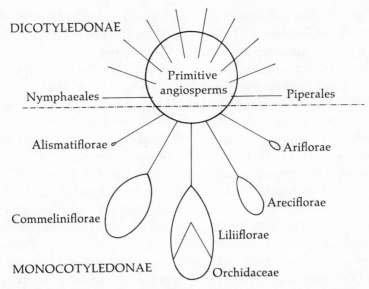

Figure 1.3 The relationships and relative sizes of the monocot superorders (very diagrammatic).

(5) Liliiflorae: a group in which inner and outer perianth segments (sepals and petals) are generally quite similar. Biochemical and other data suggest that this superorder and the Commeliniflorae are quite distinct. Liliiflorae include lilies, amaryllises, irises, and orchids.

The Orchidaceae, by far the largest family of the Liliiflorae, are generally considered to be the most highly evolved of this superorder, if not of the entire subclass Monocotyledonae. If plants were evolving toward some defined "goal," it might make sense to try to determine which group is the most highly evolved. But there are clearly as many different "directions" of evolution among plants as there are kinds of plants.

We may logically ask if any other plant group is especially closely related to the orchid family, but a clear answer is not easily found. In most plant classifications the orchids are grouped with the Burmanniaceae, Corsiaceae, and Geosiridaceae in the order Orchidales, or Microspermae. As the alternate name suggests, one of the reasons for grouping these families together is that all are characterized by tiny seeds. Tiny seeds, however, are found in all saprophytic groups, whether dicot or monocot, and are not necessarily an indication of close relationship. The presence of larger seeds with some endosperm in some members of both the Burmanniaceae and the Orchidaceae strongly indicates that tiny seeds have evolved independently in each family. In other features, the Burmanniaceae seem quite unlike the orchids. The flowers usually show radial symmetry, the stamens are united with the perianth rather than with the style, and the inflorescence is cymose. Both the Orchidaceae and the Burmanniaceae have "tenuinucellate" ovules (Huber, 1969), but the overall evidence for a close relationship between the two families seems weak. The Corsiaceae have a bilaterally symmetrical perianth, but in this family the "lip" is formed by one of the sepals rather than by a petal, so that the resemblance to the Orchidaceae is quite superficial.

Some authors have suggested a relationship between the orchids (especially the subfamily Apostasioideae) and the Hypoxidaceae, but this, too, seems to be a superficial similarity. Huber (1969) places the Hypoxidaceae in the order Asparagales and suggests that relatives of the orchids are to be sought among the "colchicoid" Liliiflorae, that is, among the Liliales, in Huber's restricted sense of the term. However, one can find no member of the modern Liliales that is especially closely related to the Orchidaceae, and Huber (1977, p. 291) says that "the Orchidales certainly did not descend from any recent member of Liliales." Though many orchids may be evolving relatively rapidly, they are by no means a young group; their lily-like ancestors probably either evolved into orchids or became extinct.

Thus, one may treat the orchids as an order, Orchidales, with only one family, the Orchidaceae. An alternative would be to include the Orchidaceae in a more broadly defined order, Liliales. Further study of the anatomy and seed structure of primitive orchids and other lilialean plants may eventually permit us to discuss the relationships of the orchid family in more definite terms.

Characteristics of the Orchid Family

Naturally, the orchids share many features with related groups of monocots: scattered vascular bundles, parallel leaf venation, flower parts in threes, an inferior ovary, and so on. We may ask, then, what the features are that distinguish orchids from all other plant groups. Surprisingly, there are very few. We can list several features that are found in most orchids, but only three that appear in all orchids, primarily because many of the evolutionary stages of the family are still found in the living members. If we had only the most primitive living representatives of the family, they might be dismissed as rather peculiar lilies. But the vast array of not-missing links show these primitive orchids to be the first stages in the evolution of this distinctive family.

The major distinguishing features of the orchid family are:

(1) The stamens are all on one side of the flower, rather than being symmetrically arranged. Most orchids have only one fertile stamen, and only one genus of living orchids has three; but whether one, two, or three, the stamens are always on one side of the flower. I believe that this simple feature was the critical step in the evolution of the orchids, but we will return to that later.

(2) The stamen and the pistil are at least partly united. In most orchids, these parts are completely united, so that we speak of a single structure, the column.

(3) The seeds are tiny and numerous. Some of the primitive orchids have much larger and more complex seeds than others, but even these must be characterized as tiny and numerous as compared with most other plants.

(4) The flower usually has a lip, or labellum. This is really the petal (or inner perianth segment) that is opposite to the fertile stamen, but it is usually so unlike the other two petals that it receives a special name. In some orchids, however, the lip is not very different from the other two petals.

(5) The flower usually twists around in the course of development. This process is known as resupination. In the orchid bud the fertile stamen develops on the side away from the stem and the lip develops on the side next to the stem. If the bud merely bent away from the stem, the lip would be uppermost and the fertile stamen would be on the underside of the flower. Some orchid flowers develop in this way and function well in this position, but most buds twist around during their development, so that the lip is on the lower side of the flower when it opens. This is so general that nonresupinate orchid flowers seem to be upside down.

(6) Part of the stigma (the rostellum) is usually involved in transferring the pollen from one flower to another. This is one of the major themes of orchid evolution.

(7) The pollen is usually bound together in a few large masses (pollinia). One of the most distinctive features of the family, this characteristic, along with the rostellum, is intimately involved in pollination by insects and birds.

Estimates of the size of the orchid family vary quite widely. Some authors suggest 12,000 or 15,000 species, others as many as 35,000. We cannot really know how many valid species exist because so few orchid groups have been well studied. The best we can hope for is a crude estimate. The numbers given in chapters 7-10, which vary from fairly accurate counts to very rough guesses, add up to 725 orchid genera and

19,192 orchid species. Considering the many uncertainties, a figure between 20,000 and 25,000 orchid species seems reasonable.

It is more difficult to guess what proportion of the family is made up of terrestrial species, or ground orchids, as compared with the epiphytic species—those that normally grow on trees. We know that about 4,000 species belong to primarily terrestrial groups, but a number of terrestrial species are to be found in groups that are primarily epiphytic. I would guess that about a quarter of the orchid species are primarily terrestrial, with perhaps another five percent being able to grow well either on the ground or in trees.

Both the Orchidaceae and the Compositae are often given as the largest family of flowering plants, and even the Leguminosae and Gramineae are sometimes proposed for the same honor, though without supporting

Figure 1.4 *Grammatophyllum papuanum*, a mature but not unusually large plant cultivated in Lae, Papua New Guinea. The stems are 2 to 3 meters long.

statistics. The best unbiased survey known to me is that of Willis (1973), which gives 17,000 species for the Orchidaceae and 13,000 for the Compositae (12,000 for the Leguminosae, 10,000 for the Gramineae). Heywood, Harborne, and Turner (1978) estimate about 22,000 species of Compositae, but we have no comparable, detailed synopsis of the Orchidaceae.

While on the subject of records, we may consider the size range of the orchids. *Vanilla* is probably the longest orchid plant, as it is a large vine and may climb many meters up a tree or cliff. The most massive orchid, however, would probably be a *Grammatophyllum* (fig. 1.4). *Grammatophyllum speciosum* and *G. papuanum* have thick stems up to nearly five meters long, and an old clump is truly massive. At the other end of the range, several species of *Bulbophyllum* have been called the world's smallest orchid, but this honor probably belongs to the Central American *Platystele jungermannioides* (fig. 1.5). Of the Old World contenders, the Bornean *Bulbophyllum odoardii* has well-developed leaves as well as pseudobulbs about at large as those of *B. minutissimum* or *B. globuliforme*. Of these two Australian species, *B. globuliforme* is thought to be smaller (Rentoul, 1977), but the pseudobulbs of *B. minutissimum*, though wider than those of *B. globuliforme*, are also much flatter. In both species the leaves are mere scales, the green pseudobulbs functioning in both storage and photosynthesis.

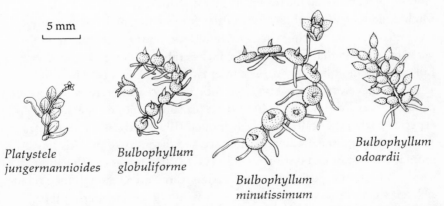

5 mm

*Platystele
jungermannioides*

*Bulbophyllum
globuliforme*

*Bulbophyllum
minutissimum*

*Bulbophyllum
odoardii*

Figure 1.5 The several contenders for the "world's smallest orchid" title, all drawn at twice life size (*Bulbophyllum globuliforme* and *B. minutissimum*, after Dockrill, 1969; *B. odoardii*, after Pfitzer, 1885).

Preface

In 1978 there appeared a two-volume work on the Compositae—a rather dull family of plants as compared with the Orchidaceae. In these volumes forty-two contributors united their efforts to review the classification, cytology, population structure, speciation, evolution, and chemistry of each composite tribe. Even with unlimited funds, I fear that we could not bring together half as many orchid specialists with enough information to build a similar review of the orchid family. Despite the many elaborate volumes with numerous color plates dedicated to orchids, there have been pitifully few careful studies on any aspect of the Orchidaceae. The botanical community—professors as well as students—are often unable to choose a reasonably well-defined and feasible thesis problem, or to distinguish between a worthwhile study and a useless one. Perhaps because of the orchids' popularity among hobbyists, most professional botanists have avoided the field, leaving a vacuum which has been filled by amateurs, some of whom have done excellent work, to be sure.

When I first worked at the Missouri Botanical Garden, I tried to learn about the flower structure of orchids. I cut flowers from as many different orchids as I could find in the greenhouses, studied these under the dissecting microscope, and tried to correlate my observations with descriptions in the literature. I soon worked out the terminology, and I learned that orchids do not always have the decency to behave as botanists say they should. I became interested in the evolution and classification of the family, and I have had these problems very much in my mind since that time. In this book, then, I will try to supply the sort of information that I had to dig out twenty years ago, and review the evolution and overall classification of the Orchidaceae. I hope the result will be of value to the hobbyist as well as to the botanist, but I must emphasize that this is not a book on how to grow orchids. There are already many such books, most of them quite good, at least for the

Geography

2

Orchids are a cosmopolitan family, ranging from northern Sweden and Alaska to Tierra del Fuego and Macquarie Island. A few vascular plants of other families are found a bit further to the north and south, but the orchids occur very near the limits of vegetation (fig. 2.1). The epiphytes among them, however, are limited to tropical and subtropical environments, as are the vascular epiphytes of most other families (some epiphytic ferns range more widely). Orchids are lacking, also, in the most extreme desert environments, though they may be found in oases, in sheltered desert canyons, and in cactus thorn shrub or thorn forest.

Like many other plant groups, the orchids are much more diverse in the tropical belt than they are to the north or south, although within the tropics they are by no means uniformly distributed (figs. 2.2, 2.3). Orchid plants may be abundant in drier forests, but such communities have relatively few orchid species. Orchids are most abundant and diverse in habitats that have over 100 inches (about 2.5 m) of annual rainfall, with no month receiving less than two or three inches (Holttum, 1960); however, really soaking-wet habitats have relatively few. Dew, or condensation from clouds, is important for some types of vegetation, especially when rainfall, as such, is limited (Blossfeld, 1974). In Central America we find the greatest abundance and diversity of orchids in wet montane forest or cloud forest from 1000 to 2000 meters in elevation. A similar situation occurs in the South American Andes, though the orchid-rich zone may be somewhat higher. Vareschi (1976), comparing the number of orchid species per hectare in Venezuela with an index of diversity (based on number of species and leaf diversity) for the vegetation as a whole, found comparatively more species in tepui forest than in montane or cloud forest (fig. 2.4). While these results are interesting, one feels that the sampling may not be adequate. Dunsterville (1961), for example, has found at least 47 different species of epiphytic orchids on

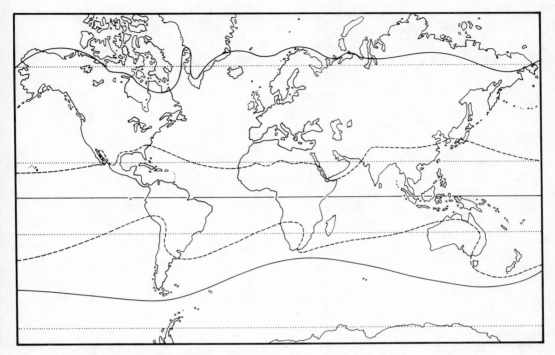

Figure 2.1   The approximate northern and southern limits of terrestrial orchids (solid lines) and the approximate limits of epiphytic orchids (dashed lines).

one tree in Venezuelan "montane rain forest (perhaps even justifiably rated as cloud forest)." I have found 60 to 100 different species in small areas of similar habitat in Panama, though no attempt was made to calculate the number of species per hectare. It is much easier to find all the orchid species present in an elfin forest or in the "tepui" forest than it is in montane forest or cloud forest, where the trees are much larger. Montane forest and cloud forest are difficult to sample adequately without cutting the trees.

The most striking thing about the distribution of orchids is that the different continents have distinctive orchid floras. In other words, a large part of orchid evolution has occurred since the continents became well separated. We will consider this in somewhat greater detail below.

Dispersal

With the exception of a few of the most primitive genera, orchids have seeds that are well suited for wind dispersal. The exceptions—*Apostasia, Selenipedium,* and *Vanilla*—have small but relatively heavy seeds that are borne in a somewhat pulpy fruit. It may be that the aroma of *Seleni-*

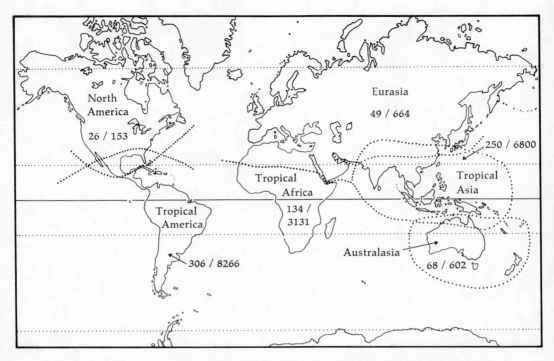

Figure 2.2 The major geographic regions used in this book. Adjacent regions overlap except when separated by ocean or desert barriers. The approximate numbers of orchid genera and species in each region are given (genera / species).

pedium and *Vanilla* serves to attract insects or other animals that eat the pulp and disperse the seeds, but we have no direct knowledge of how the seeds are normally dispersed. Dispersal by water may occur in a few cases, such as *Epipactis gigantea*. But most orchids, with their tiny, dust-like seeds, seem well suited for long-distance dispersal by wind. This subject always arouses great controversy. Some insist that long-distance dispersal is very important, and others say that they have never seen it; it doesn't happen. The truth surely lies somewhere in the middle. Most orchid seeds fall within a few meters of the parent plant, but some are carried many kilometers away. The probability decreases with distance, but in geological time, even a tiny probability becomes respectable. Orchid seeds may survive for long periods if dessicated and cool (Sanford, 1974); so we cannot discount the likelihood that one in ten billion orchid seeds is blown from Africa to South America in viable condition. Dispersal, however, is only the first step. In order for a seed to germinate in a new area, it must find physical conditions that permit germination and an appropriate fungus with which it can form a mycorrhiza. Even

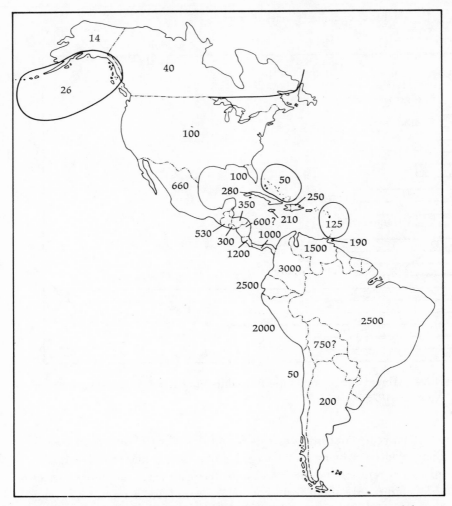

Figure 2.3 Approximate numbers of orchid species in different regions of the western hemisphere. The northern areas are partly delimited on ecological rather than national boundaries, to emphasize the progression from the arctic to more favorable habitats. Though the estimates are higher than the numbers recorded in several cases, they are probably conservative.

then, in order to become established, the mature plant must be either self-pollinating or self-compatible, and, if the latter, must find a pollinator enough like its normal pollinator to ensure a few seed capsules in the first few generations of the pioneer species.

One of the finest examples of dispersal in action occurred on the island of Krakatoa (Doctors van Leeuwen, 1936). In 1883 a volcanic explosion and the subsequent rain of hot cinders destroyed all life there.

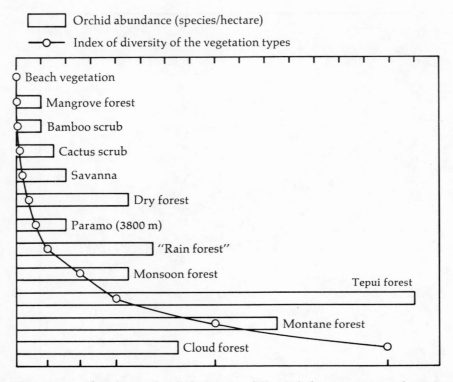

Orchid abundance (species/hectare)

Index of diversity of the vegetation types

Beach vegetation

Mangrove forest

Bamboo scrub

Cactus scrub

Savanna

Dry forest

Paramo (3800 m)

"Rain forest"

Monsoon forest

Tepui forest

Montane forest

Cloud forest

Figure 2.4 Abundance of orchid species in different habitats in Venezuela. (After Vareschi, 1976.)

Thirteen years later investigators found *Arundina, Cymbidium,* and *Spathoglottis* growing on the island, and by 1933 they found eighteen species of terrestrial orchids and seventeen species of epiphytes. Because the island is only about forty kilometers from Java, dispersal was relatively rapid. In the case of the extremely isolated Hawaiian archipelago, only three orchid species reached the islands and became established in prehistoric times, compared with about 270 other vascular plant species that managed to do so. Now, of course, other orchid species have escaped and are escaping from gardens and finding a place in the island ecology.

Geography through Time

Only twenty-five years ago most biologists believed that the continents, as we know them, had remained essentially unchanged through geological time. The idea of continental drift had been suggested much earlier but was usually dismissed either as a crackpot idea or as something that occurred too long ago to influence modern plant or animal distributions. Some authors tried to explain plant and animal distributions by hypothetical "land bridges" between continents, and the number of such bridges that were proposed for different geological periods became quite absurd. At the same time, other writers insisted that most disjunctions arose through long-distance dispersal and local extinction. Now geologists have presented overwhelming evidence in favor of continental drift, and there is no doubt that we must consider past geography in order to understand the distribution of present-day plants and animals.

The ancient continent of Gondwanaland was already breaking up in the early Cretaceous period, when the angiosperms first appeared in the fossil record, but the pieces were closer together and differently arranged than are the modern continents (fig. 2.5). In the early stages of flowering plant evolution, the major areas with a tropical climate were closer together, and dispersal between them should have been much easier than at present. Since then, they have become more and more widely separated, and direct dispersal between South America, Africa, and tropical Asia has been severely limited during much of the Tertiary period (fig. 2.6). Dispersal between Eurasia and North America, on the other hand, has been relatively easy during much of the Tertiary. One must remember, also, that Antarctica was once much warmer, permitting some dis-

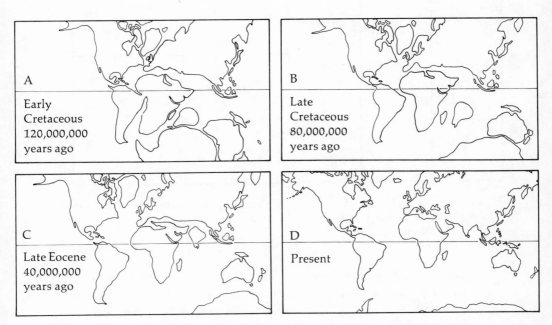

Figure 2.5 The relative positions of the continents in the early and late Cretaceous periods, the late Eocene epoch, and the present. (Modified from Smith and Briden, 1977.)

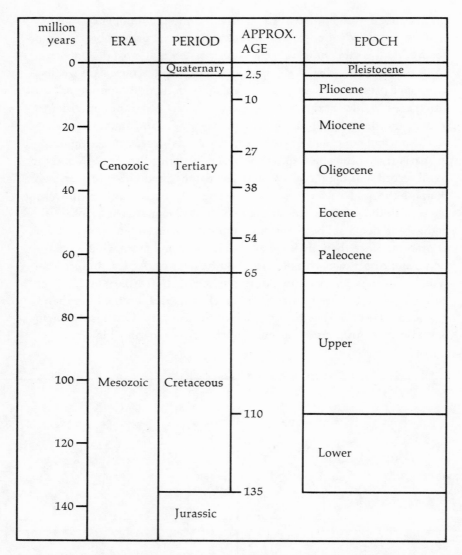

million years	ERA	PERIOD	APPROX. AGE	EPOCH
0		Quaternary	2.5	Pleistocene
				Pliocene
			10	
20	Cenozoic	Tertiary		Miocene
			27	
				Oligocene
40			38	
				Eocene
			54	
60				Paleocene
			65	
80				Upper
100	Mesozoic	Cretaceous		
			110	
120				Lower
			135	
140		Jurassic		

Figure 2.6 The geological time scale for the last 140 million years, during the evolution of the flowering plants. (From Raven and Axelrod, 1974.)

persal between southern South America and Australia and New Zealand.

Climatic changes through time are nearly as important as changes in land distribution in the history of the world's vegetation. We have a great deal of information on climatic variation in the last two million years, but this has been, at best, an atypical epoch. During most of geological time the world's climate has been more equable, with no ice cap at either pole. Thus, the polar areas bore cool-temperate vegetation, and both tropical and subtropical vegetation zones spread further from the equator than is now the case. Climates were apparently becoming somewhat warmer during the Cretaceous period, with the maximum temperature occurring possibly as late as the Eocene epoch. The climate gradually cooled during the rest of the Tertiary. The Pleistocene epoch was characterized by climatic fluctuation, its four major glaciations alternating with periods as warm as the present or warmer. At the same time there was fluctuation in rainfall and temperature in many parts of the earth, with unusually high mountain ranges also contributing to the climatic contrasts. The Pleistocene climatic changes were especially drastic in Europe, where mountains and drier areas prevented the southern migration of many species. Many of the warm-temperate plants now found in North America and eastern Asia occurred in Europe in the mid-Tertiary but became extinct there in the late Tertiary or the Pleistocene. Similarily, the African vegetation has suffered great changes during the late Tertiary and Pleistocene and is now relatively poor as compared with that of other tropical areas.

Disjunctions

Because the three main tropical regions shown in figure 2.2 have, for the most part, different orchid floras, those few genera that occur on two or more isolated continents may give us clues about the dispersal and evolution of the orchids. We will not consider genera that have nearly continuous distributions around the Northern Hemisphere, such as *Cypripedium, Listera,* or *Spiranthes,* though these each occur on two or more continents. Garay (1964) finds that only 32 or 33 of about 800 orchid genera show transoceanic distributions, and that 27 of these are terrestrial. Garay has grouped these genera according to geographic pattern (table 2.1), but genera which show a similar geographic pattern do not necessarily have the same history. For this reason I have attempted to divide the disjunct genera into three age groups as well. These groupings, however, are only segments in a spectrum, and we cannot hope to draw sharp lines between them without a fossil record.

(1) Old dispersals (figs. 2.7, 2.8, 2.9): All these are quite primitive genera (or pairs of closely related genera) with transoceanic tropical distributions; the species in different continents are well differentiated. These disjunctions probably date back to the very early Tertiary period, when the present-day continents were much closer together.

(2) Middle-aged hoppers: These genera probably represent long-distance dispersal in the mid Tertiary. In each case, considerable specia-

Table 2.1 The orchid genera which show disjunct, transoceanic distributions, arranged according to geographic pattern (modified from Garay, 1964), and by probable age of dispersal.

Age of Dispersal	Pantropical	Africa, South America	Asia, Africa	Transpacific	Asia, North America
Old dispersals	Corymborkis Vanilla	Palmorchis- Diceratostele		Epistephium- Clematepistephium Tropidia	
Middle-aged hoppers	Bulbophyllum Calanthe Eulophia Goodyera Habenaria Liparis Malaxis Polystachya		Brachycorythis Cheirostylis Disperis Nervilia Phaius Satyrium Zeuxine	Cranichis- Coilochilus Erythrodes	Aplectrum Pogonia Tipularia
Young hoppers	Polystachya concreta	Oecoclades Pteroglossaspis	Acampe Agrostophyllum Angraecum Galeola(?) Oberonia Taeniophyllum		

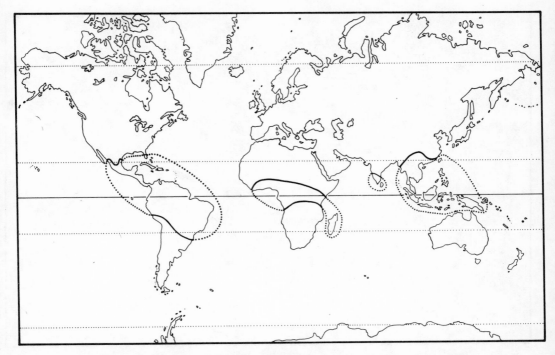

Figure 2.7 The distribution of the genus *Vanilla*. (Data from Portères, in Bouriquet, 1954.)

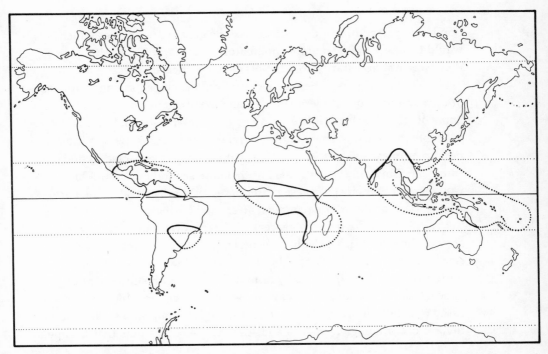

Figure 2.8 The distribution of the genus *Corymborkis*. (Data from Rasmussen, 1977.)

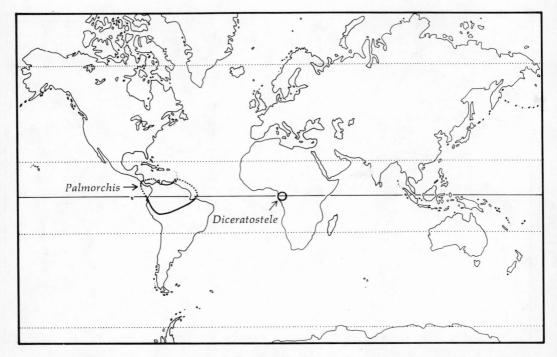

Figure 2.9 The distribution of *Diceratostele* and *Palmorchis*, a pair of closely related genera.

tion has occurred in two or more areas. The genera are all ones which would be likely to find suitable pollinators in any tropical habitat, and they were not necessarily contemporaneous in their dispersal.

(3) Young hoppers: These genera are well developed in one continent or area and have one or very few disjunct members in a second area. Some of these, such as *Galeola*, may be older distributions, but it is probable that most of them represent relatively recent dispersals into a new area. In each case, they are genera that could be expected to find suitable pollinators (though some of them are self-pollinating). One would guess that most of these patterns represent dispersal in the very late Tertiary or Pleistocene, or even after the Pleistocene.

In addition to the truly transoceanic patterns, we find relationships around the northern hemisphere, especially between eastern Asia and southeastern North America. These genera are thought to represent northern migrations in the early or mid Tertiary, with subsequent disappearance in intermediate areas. In the past, when most biologists refused to believe in the mobility of continents, this migration was thought to have occurred between Asia and Alaska. We realize now that Europe

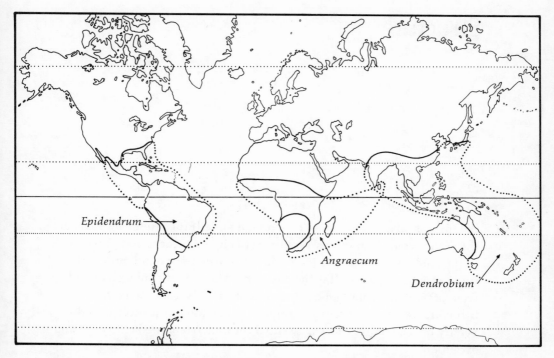

Figure 2.10 The distribution of *Epidendrum, Angraecum,* and *Dendrobium,* each a genus characteristic of one area. The genus *Angraecum* has an outlying species on the island of Ceylon *(A. zeylanicum)*.

and America were connected during the early Tertiary, and that migration by island hopping was relatively easy for some time after the land connection was broken (McKenna, 1975).

Structure

3

Botany has, alas, a reputation (often undeserved) for producing some of the dullest courses in the university curriculum. Of these, morphology, or the study of structure, is too often the most boring of all. Still, structure cannot be ignored. It is the means by which we distinguish orchids from other plants, and different groups of orchids from each other. Structural details are necessary for any discussion of the ecological diversity of the family, and are essential for understanding pollination and evolution.

Growth Habits

We have already mentioned the rather fundamental difference in growth between monocots and dicots. Dicots normally have a vascular cambium which allows the stems to grow in diameter as needed during the entire life of the plant. Most monocots lack such a tissue, though a few (none orchids) have something similar. Thus, once a section of orchid stem has matured, it cannot become any thicker through growth. In discussing stem structure, we speak of nodes and internodes, the nodes being the points at which the leaves are attached, an internode the section of stem between two such points (fig. 3.1B). Some botanists like to consider plants as made up of phytomeres, or units made up of one internode, one leaf (more in some other plants), and an axillary bud, which has the ability to make new units. Among the monocots these units are not obscured by secondary growth, though the stem can produce roots at the nodes. Further, the monocots usually show intercalary growth; that is, the basal part of the internode retains growth potential and continues to elongate for some time, supported by the leaf sheaths that surround it. This is obvious in orchids that flower from the apex of a pseudobulb before the pseudobulb itself has matured, as in many species of *Coelogyne*. Once a stem has matured, however, new growth (except for

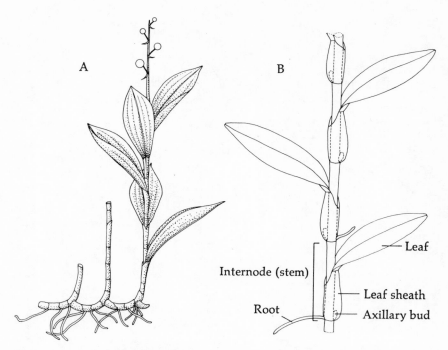

Figure 3.1 *(A)* The growth habit of a primitive orchid, with spirally arranged, plicate leaves and a terminal inflorescence. Each year a new shoot is produced from the base of a previous shoot. The basal portions of successive shoots together form the horizontal rhizome. *(B)* A diagrammatic sketch of three internodes of a monocot stem. The basic unit, an internode with a leaf and an axillary bud (and the potential of producing roots), is repeated. Changes in the proportion of the stem and leaf and the presence or absence of roots account for much of the variation in plant form.

roots) normally occurs only from an apical bud or an axillary bud. A bud may form a new shoot, an inflorescence, or a single flower, depending on its position and the growth habit of the plant.

Holttum (1955a) has suggested that the basic growth form of the monocots is sympodial, with each shoot having definitely limited growth and being succeeded by a similar shoot from an axillary bud (fig. 3.1A). The monopodial growth habit, in which the apex of a shoot has the potential for unlimited growth, has been derived from the sympodial habit in various groups of orchids (fig. 3.2N–P). Sympodial plants may have shoots clustered together or spread out on a long rhizome, and new shoots may arise from any part of the older shoots where there is an axillary bud (fig. 3.3, A–M). Most growth forms, whether sympodial or monopodial, may be erect, creeping, or pendant. An interesting varia-

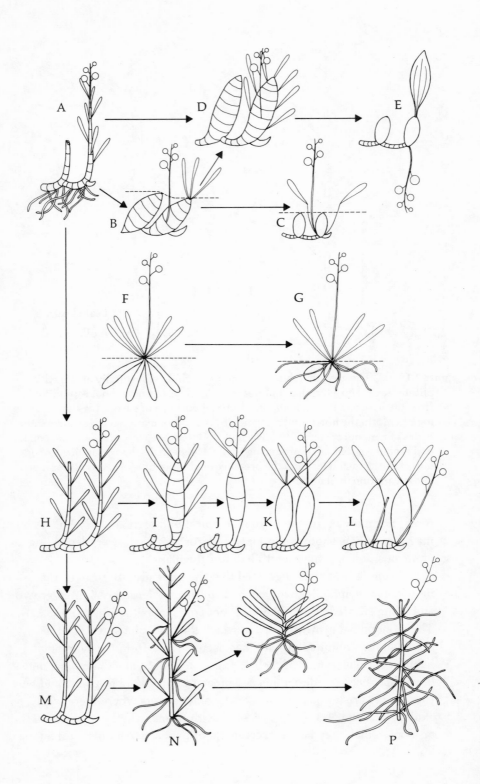

tion is found in *Oncidium* section *Serpentia*, which taxonomists describe as having a "long, wiry rhizome." Daniels and Rodríguez (1972) have shown that this is actually a twining inflorescence, up to five meters long, which produces plantlets at the nodes. These inflorescences often make a tangled mat in the treetops.

Roots

Orchids, being monocots, never have a carrot-like taproot, and, indeed, never have a primary root like that of most dicots. The entire root system is made up of secondary roots that arise from the stem. Although they vary greatly in thickness, orchid roots are never as thin and fibrous as those of grasses and some other monocots.

One of the features found in most orchid roots is the velamen or, in the older literature, the *velamen radicum*. This is an outer layer of cells with partially thickened cell walls which lose their living contents as the root matures (fig. 3.4). In the simplest case, it is a single layer of cells, equivalent to the epidermis in all respects. In many cases, however, there are from two to eighteen layers of cells. Some have defined the velamen as a many-layered epidermis (though others define the epidermis as the outermost layer of cells). The velamen is usually a spongy, whitish sheath around the root, and between the velamen and the cortex is a special layer of cells, the exodermis. This includes long, thick-walled cells which lack cytoplasm and shorter, living, "passage" cells. There has been some controversy over the function of the velamen, but Capesius and Barthlott (1975) and Barthlott (1976a) have shown that the aerial roots do absorb water and nutrients through the velamen (except in *Vanilla*). Having a spongy, water-absorbing layer around the root seems clearly adaptive for epiphytes. Went (1940) suggests that the great value of the velamen lies in its ability to catch and hold the first, relatively mineral-rich water to reach the roots whenever it rains. A number of authors mention finding algae within the velamen, and some specify blue-green algae. One wonders if these may not be encouraged guests

Figure 3.2 Different plant habits in the orchid family, showing probable patterns of evolution (highly diagrammatic). The growth habit is sympodial in *A–M*, monopodial in *N–P*. The leaves are spiral in *A*, distichous (arranged in two rows) in all others. The inflorescence is terminal in *A* and *F–K*, lateral in all others. Corms of several internodes are shown in *B*, of a single internode in *C*. Pseudobulbs of several internodes are shown in *D, I,* and *J*; pseudobulbs of single internodes are shown in *E, K,* and *L*. Fleshy storage roots (tuberoids) the shown in *F*, and root-stem tuberoids in *G* (both terrestrial). A leafless plant with photosynthetic roots is shown in *P*.

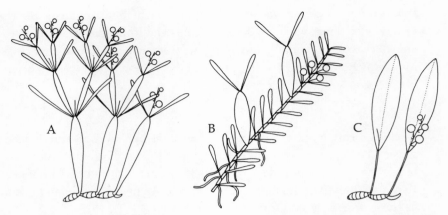

Figure 3.3 Unusual habits. (A) A sympodial plant in which each pseudobulb produces two new pseudobulbs apically after flowering, as in *Scaphyglottis*. (B) A
plant with pseudobulbs scattered on an elongate, leafy stem, as in some
Maxillaria species. (C) A sympodial plant with no pseudobulbs but a fleshy leaf,
as in most Pleurothallidinae.

that fix nitrogen in the velamen. The velamen cells commonly contain
mycorrhizal fungi, and the origin of the tissue may be, in some way,
related to the mycorrhizal relationship.

The velamen is most obvious in epiphytes, and especially in bark and
twig epiphytes, but it is also present in most terrestrials, as well as in
some Liliaceae and Araceae (Mulay and Deshpande, 1959; Mulay, Deshpande, and Williams, 1958). In epiphytic orchids the growing apex of
the root is commonly green when exposed to light, and there are often
green chloroplasts within the cortex of the mature root, though the velamen masks their green color. In some Vandeae, all photosynthesis takes
place in the roots, the stem being very short and bearing only scale-
leaves. In one of these leafless Vandeae, the genus *Taeniophyllum*, the
outer velamen cells show a curious pattern, many of the cells having
large, regular holes in the outer walls, with clusters or islands of cells
which lack these openings (fig. 3.5).

In many terrestrial orchids the roots form storage organs, or tuberoids
(a tuber is, by definition, a stem). In some, such as the Spiranthinae, the
whole root is very fleshy. In other cases, as in *Cleistes* or some species of
Tropidia, some roots or parts of roots are thick and others much thinner.
We may call these nodular tuberoids, to distinguish them from roots that
are thick throughout. In some Goodyerinae, such as *Cheirostylis*, the
fleshy rhizome does not bear roots as such, but only "root ridges"
(*Wurzelschwielen*; Burgeff, 1932). Each ridge has root anatomy, with

A

B

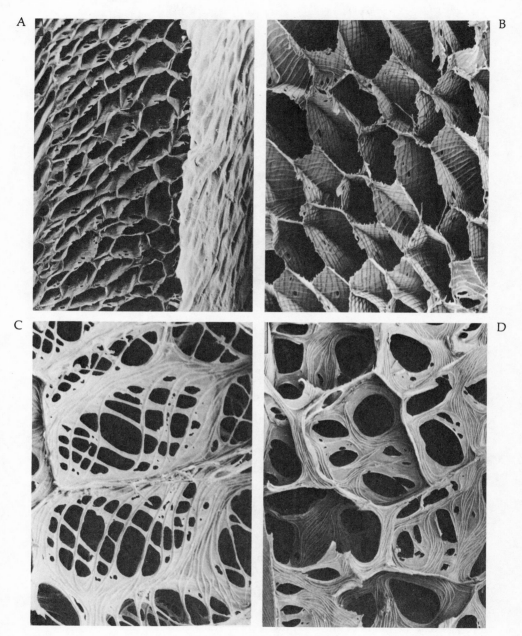

C

D

Figure 3.4 Scanning electron microscope photographs of orchid velamen. *(A)* Section through the root of *Vanda tricolor,* showing the surface of the root and the velamen in section. *(B) Dendrobium superbum,* section through velamen. *(C) Graphorkis lurida,* section through velamen. *(D) Clowesia russelliana,* section through velamen. (Courtesy of W. Barthlott.)

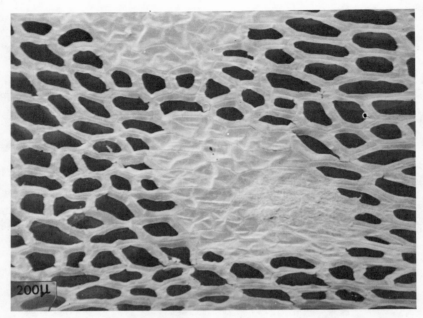

Figure 3.5 Surface of the velamen in *Taeniophyllum* species. (Courtesy of W. Barthlott.)

root hairs on the outer surface, and has a root stele, or vascular cylinder, running from the vascular system of the rhizome along its length (fig. 3.6). Another most unusual structure is the root-stem tuberoid of the tribes Orchideae, Diseae, and Diurideae (fig. 3.7). It is largely a storage root, but the basal portion has a sheath of root structure around a core of stem structure with an apical bud. This is the structure that survives during the dormant season; in the growing season the bud grows into a new shoot, with one of the axillary buds forming a new tuberoid, which will be the next resting phase. This structure is also known as the "dropper" or "sinker," as in some cases it may, with an extension of the leaf base, grow deeper into the soil. In both the Orchideae and the Diurideae this structure is often polystelic, that is, it includes several vascular cylinders, as though several roots were grown together in one skin.

Most members of the Orchideae have spheroid tuberoids. During the growing season the plant typically has two of these structures, an old one from the previous season and a new one that will survive during the coming dormant season. These paired structures look rather like mammalian testicles, and the plants were thus called *orchis*, or testicle, by the

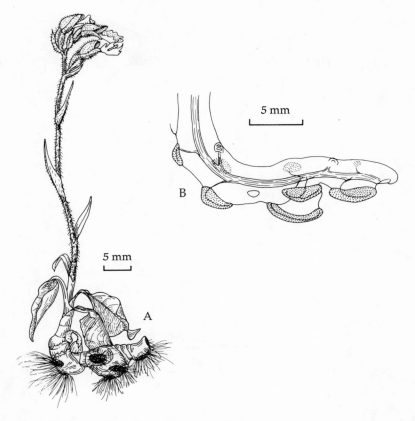

Figure 3.6 *(A)* Growth habit of *Cheirostylis philippinensis*, showing the root ridges on the fleshy rhizome. *(B)* A cleared section of the rhizome, showing the vascular tissue and the ridges with root anatomy (stippled). (After Burgeff, 1932.)

ancient Greeks. This has become the generic name of a European genus, and has given the family its name.

With reference to epiphytic orchids, growers often speak of aerial roots, which do not take kindly to being covered up in a pot, and substrate roots, which enter the potting medium. These types of root are recognizably different, but reviews of orchid anatomy are strangely silent on their structure. Roots that are wholly aerial or completely within the substrate are usually cylindrical, but roots which creep on the surface are often somewhat flattened or dorsoventral. In some Vandeae, such as *Phalaenopsis* and *Dendrophylax*, the roots are strongly flattened, whether creeping or not. Fleshy storage roots are not limited to the terrestrials. Epiphytes and lithophytes, such as *Sobralia, Ponera, Isochilus* and some species of *Epidendrum*, may have very thick, fleshy roots.

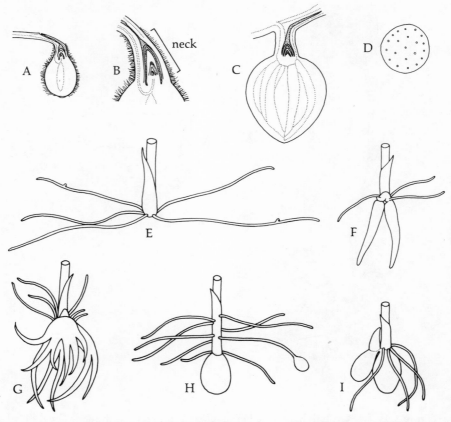

Figure 3.7 Structure of the root-stem tuberoid. (*A*) Longitudinal section through a root-stem tuberoid of *Pecteilis*. (*B*) Part of the same at higher magnification, showing the "neck" that has a sheath of root tissue around a core of stem tissue. (*C*) Longitudinal section through a root-stem tuberoid of *Orchis morio*, showing the arrangement of the many steles. (*D*) Cross-section of the root-stem tuberoid of *Platanthera*, showing the many steles. (*E–I*) Different forms of root-stem tuberoid: (*E*) stoloniferous tuberoids (*Platanthera*); (*F*) fusiform tuberoids (*Platanthera*); (*G*) palmate tuberoids (*Gymnadenia*); (*H*) stalked spheroid tuberoids (*Pecteilis*); (*I*) spheroid tuberoids without stalks (*Orchis*). (*A, B* after Kumazawa, 1956; *C* after Stojanow, 1916; *D–I* after Ogura, 1953.)

Plants of this sort generally have thin stems and leaves, and many a collector has learned, to his sorrow, that the plants can survive with their shoots cut off, but not without roots.

Some genera show a special sort of aerial root. In *Ansellia, Cyrtopodium, Grammatophyllum,* and a few other genera there are, in addition to normal roots, nonabsorbing roots that tend to project upward from the substrate (Barthlott, 1976a). These may function in the accumulation of debris over the normal, absorbing roots (see Chapter 4).

Roots normally produce only roots, but "adventitious" buds, which give rise to new shoots, may be produced on roots, as in *Listera, Pogonia, Psilochilus,* or *Phalaenopsis* (Stoutamire, 1974b). Such roots are superficially much like slender root-stem tuberoids, but they are quite different in structure.

Stems

Basically, the orchid stem is much like that of corn or lilies, or any other ordinary monocot stem. The vascular tissue is in many scattered bundles, these usually being denser toward the periphery of the stem and embedded in softer parenchyma, or storage tissue. The stems may be thin and wiry, somewhat woody, or soft and succulent, as in *Vanilla*.

RHIZOME

The term "rhizome" is used for any horizontal stem on, or in, the substrate. In the majority of sympodial orchids the rhizome is a compound organ, made up of the basal sections of successive shoots. Typically, the first few internodes of the shoot are horizontal, somewhat thickened, and quite hard and woody. The axillary buds at a few of the nodes are unusually well developed, and one or two of these grow to form the next shoot(s). In sympodial orchids with lateral inflorescence, the inflorescence also may arise from the rhizome. "Rhizome" is also used for the creeping, horizontal stems of the Goodyerinae, though there is usually not a sharp distinction between the rhizome and the aerial shoot in that group. In orchids with monopodial habit of growth there is no distinct rhizome.

The term "secondary stem" which is found in many taxonomic descriptions seems to refer to the vegetative shoot above the rhizome. This use is inaccurate and confusing. Morphologically, the only primary stem is the first seedling shoot (from a protocorm). All other stems are secondary, including the rhizome. "Aerial stems" or "erect vegetative shoots," although awkward, are accurate enough terms in the case of terrestrials. In epiphytes, the entire shoot is aerial, and in many cases the vegetative shoots may be horizontal or pendant. Unfortunately, much of our terminology has developed more by accident than by plan, and I cannot suggest a really good substitute for the inappropriate "secondary stem."

CORMS AND PSEUDOBULBS

The term "corm" is used to refer to an underground storage stem like that of *Gladiolus.* Thus, *Bletia, Eulophia, Spathoglottis,* and some related genera have quite typical corms. There is no sharp line, however, between corm and pseudobulb. The term "bulb," as an English botanical term, refers only to a structure like that of the tulip or the onion, in which the bulb is largely made up of thickened leaf bases. The term "pseudobulb" was applied to the thickened stem structures of epiphytic orchids, which are neither bulb nor tuber, and the term has achieved general usage. A pseudobulb may be made up of a single thickened internode (heteroblastic) or several (homoblastic), and a pseudobulb of several internodes may bear leaves along its length or only at the apex. In a few orchids the pseudobulb is surrounded by leaf-bearing sheaths, but bears only a scale leaf at its apex. Pseudobulbs are quite incompatible with the monopodial growth habit, and so monopodial orchids must use either leaves or roots as storage organs.

A curious cavity is found in the apex of the pseudobulb of the dwarf *Bulbophyllum minutissimum* and *B. odoardii* (Pfitzer, 1884). In each case this cavity has a small opening to the outside; in *B. minutissimum* the opening is partially concealed by the scale-like leaf or leaves. The cavity contains stomata through which the tissues of the pseudobulb can exchange gases with the atmosphere. *Bulbophyllum odoardii* also has stomata on the leaf; the pseudobulb presumably continues to function as a photosynthetic organ after the leaf has fallen off.

Eria bractescens and its allies have a different sort of cavity (Kerr, 1971). In these orchids the inflorescence bud is formed at the base of a cylindrical hole in the pseudobulb. This hole is covered by a leaf sheath, and these sheaths are firmly attached to the surface of the pseudobulb. When the inflorescence develops, it breaks through the sheath, sometimes forming a "trapdoor." When the old inflorescence rots away, it leaves a neat, cylindrical hole in the pseudobulb.

Leaves

As we have noted, each node of the stem bears some sort of leaf-like organ, with an axillary bud at its base. In many orchids the rhizome bears mere scale leaves or leaf sheaths. Most orchid leaves are typical of the monocots, with many parallel veins, the cross connections between the longitudinal veins being inconspicuous. In *Epistephium* and *Clematepistephium*, however, it is hard to call the leaf anything but net-veined (fig. 3.8A). The arrow-shaped leaves of *Pachyplectron* (B) also look a bit out of place among the orchids, and the deeply lobed leaves of some *Acianthus* species (F) would seem more at home on a buttercup.

ARRANGEMENT

In the majority of orchids the leaves are arranged distichously, or in two ranks, with the leaves alternating on opposite sides of the stem (fig. 3.2

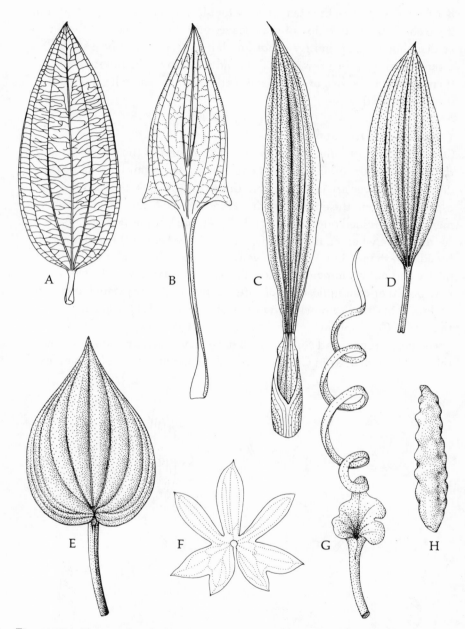

Figure 3.8   Various types of orchid leaf. *(A) Clematepistephium*, net-veined. *(B)*
Pachyplectron arifolium, net-veined and hastate. *(C) Catasetum*, plicate with a
sheathing base. *(D) Stanhopea*, plicate with a distinct petiole. *(E) Monophyll-
lorchis maculata*, plicate and cordate. *(F) Acianthus bracteatus*, conduplicate and
deeply lobed. *(G) Thelymitra spiralis*, twisted. *(H) Dendrobium cucumerinum*,
fleshy.

B–O). In many cases the stem or pseudobulb bears only a single leaf, but if we check the orientation of scale leaves and sheaths, we find them to be distichous. The primitive condition, however, seems to be a spiral arrangement *(A)*. In a few cases, by condensation of the internodes, there appear to be two or more leaves arising at the same level *(Codonorchis, Isotria)*.

VERNATION AND FOLDING

One of the features much used in orchid taxonomy is the way in which the leaves fold or roll in development. Among the primitive groups the developing leaves are rolled, or convolute (fig. 3.9*A,B*). In many, especially among the epiphytes, the leaves are duplicate during development *(C)*, or folded once with each half flat. Such leaves are always conduplicate when mature, that is, with a single fold at the midline and broadly V-shaped in section, the veins of the leaf blade all being similar in size and not prominent *(E)*. Leaves that are convolute in development may be either conduplicate or plicate. In the plicate, or pleated, condition several veins are prominent, and the leaf is usually folded at each of these *(D)*. Conduplicate leaves have evolved from plicate leaves in several orchid groups, and some groups have leaves that are not quite one thing or the other, as in *Cymbidium* and the *Chondrorhyncha* complex.

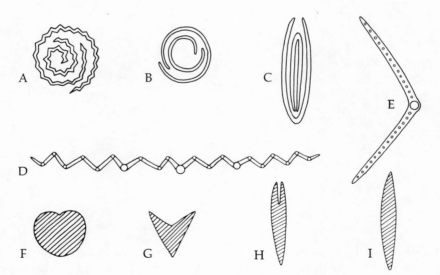

Figure 3.9 Diagrammatic cross sections of developing and mature leaves. *(A, B)* Convolute development. *(C)* Duplicate development. *(D)* Plicate leaf. *(E)* Conduplicate leaf. *(F)* Cylindrical leaf. *(G)* Triangular leaf. *(H, I)* Laterally flattened leaf at different levels.

Bletia, Sobralia, and *Spathoglottis* are genera with typically plicate leaves. Plicate leaves are always relatively thin, as are, usually, the conduplicate leaves with convolute development, but the conduplicate leaves with duplicate vernation may be either thin or extremely fleshy. In the extreme case, fleshy leaves may be triangular or cylindrical in form *(F,G).* In a number of orchids we find the leaves to be laterally flattened, rather than dorsoventral *(H,I).* These are often called "equitant," as each leaf "rides" on the one above it.

SHEATHS AND PETIOLES
In many cases, the basal portion of the leaf forms a sheath around the stem, a feature without which there could be little intercalary growth, the soft, meristematic region of the internode being too soft to support itself alone. (fig. 3.10*A*). In other cases such sheaths are formed without leaf blades *(C),* and especially on rhizomes and inflorescences the sheaths may be small. In the genus *Teuscheria* the spotted sheath about the pseudobulb is hard and maintains its form even after the pseudobulb has shriveled within. In some orchids the base of the leaf forms a narrow, subcylindric petiole, as in some Stanhopeinae (fig. 3.8).

ARTICULATION
We usually take the falling of leaves for granted, especially in autumn,

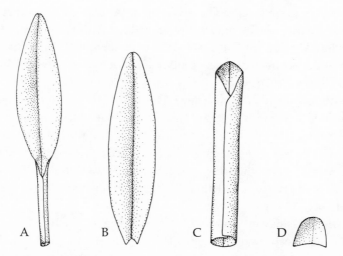

Figure 3.10   Diagrams of different leaf-life structures which may occur on a single plant. *(A)* Leaf with sheathing base. *(B)* Leaf without sheathing base. *(C)* Sheath without leaf blade. *(D)* Scale-like bract.

but not all leaves are capable of breaking away. A special abscission layer of breakaway cells near the base of the leaf, commonly called a joint or articulation, is lacking in many terrestrial orchids, and so the leaves simply rot in place. While an abscission layer is quite lacking in four of the six subfamilies, it is found in most members of the Epidendroideae and Vandoideae, though it may be lost again in some highly evolved micro-orchids such as some species of *Dichaea*, *Epidendrum*, and *Notylia*. In most cases the abscission layer is between the leaf sheath, if there is one, and the blade. Where no leaf sheath is found, the leaf commonly abscisses at the base. In a few cases, such as *Teuscheria* and some *Oecoclades* species, the abscission layer is well above the base of the petiole. One might, at first glance, suspect the structure below the joint to be the apex of the pseudobulb, but in some species of *Oecoclades* there are two leaves on a single pseudobulb, each with a joint in the middle of the petiole (Summerhayes, 1957).

STOMATA AND SUBSIDIARY CELLS

Land plants normally have small openings, or stomata, on one or both leaf surfaces which permit the passage of gases, and especially the entry of carbon dioxide into the leaf. Each opening is bordered by two kidney-shaped "guard cells" which can open and close the stomata by changing their shape. The guard cells are very different from other epidermal cells. In many cases the stomata and guard cells are accompanied by two or more distinctive subsidiary cells that are also structurally unlike the surrounding epidermal cells. The presence or absence of subsidiary cells, and especially their developmental relationships with the guard cells, are considered important features in plant classification. As far as is now known, there are three patterns of stomatal development in the Orchidaceae (Williams, 1979):

(1) Without subsidiary cells (fig. 3.11*A–C*): all Orchidoideae, some Cypripedioideae, some Apostasioideae, Pogoniinae, and *Epistephium*.

(2) Mesoperigenous subsidiary cells *(D)*: One of the subsidiary cells is derived from the same meristemoid as the guard cells; the other is derived from another neighboring cell; found in Spiranthoideae. This is the only known occurrence of this developmental pattern in the monocots (Williams, 1975).

(3) Perigenous development with trapezoid cells *(E–F)*: in this pattern of development, cells on either side of the meristemoid (each in a different row of cells) produce trapezoid cells by oblique cell divisions. The trapezoid cell may develop into the subsidiary cell or it may divide to produce one or two subsidiary cells. There may also be "polar" subsidi-

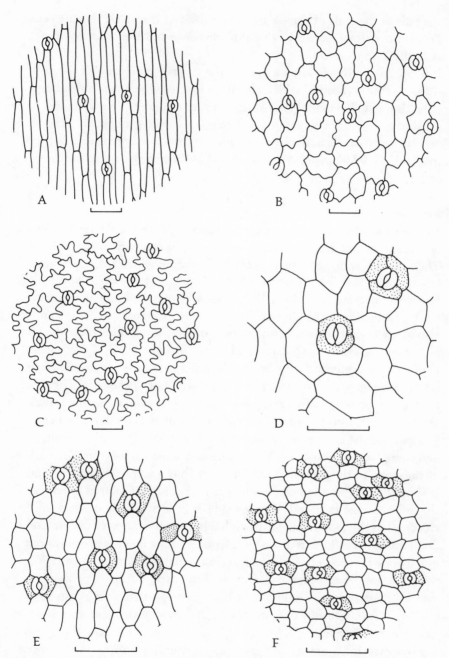

Figure 3.11 The patterns of epidermal cells from the undersides of different orchid leaves; subsidiary cells stippled. *(A) Calochilus herbaceus* (Diurideae). *(B) Habenaria petalodes* (Orchideae). *(C) Isotria verticillata* (Vanilleae). *(D) Spiranthes (=Beadlea) elata* (Cranichideae). *(E) Sobralia fragrans* (Arethuseae). *(F) Neomoorea irrorata* (Maxillarieae). Subsidiary cells are lacking in *A, B,* and *C,* mesoperigynous in *D,* and perigenous (trapezoid) in *E* and *F*. Scale 0.1 mm. **(Courtesy of N. H. Williams.)**

ary cells at the ends of the stomata, also derived from neighboring cells. This is the pattern in nearly all Epidendroideae and Vandoideae, as well as in the Triphoreae.

Subsidiary cells are known in both the Cypripedioideae and the Apostasioideae (both of which also have species which lack subsidiary cells), but developmental studies are not yet available. Four or more subsidiary cells are characteristic of highly evolved groups. Patterns with two subsidiary cells and with no subsidiary cells are both found in primitive orchids. Williams (1979) suggests that the absence of subsidiary cells is primitive for the family.

Inflorescence

The orchid inflorescence is normally racemose, with the flowers axillary on the rachis and usually flowering from the base upwards (fig. 3.12A). There are a few orchids, such as *Orchis simia*, in which the uppermost flowers open first, but the inflorescence is still a raceme even though the order of flowering is modified. Inflorescences may be branched (paniculate), in which case the ultimate branches are racemose (B). There are also many orchids which normally bear one-flowered inflorescenses, such as *Lycaste*, *Maxillaria*, *Dichaea*, and *Chondrorhyncha* (D). In all cases, the flower is subtended by a bract. The bract is usually inconspicuous, but may be large or colored, as in *Cyrtopodium* or *Lockhartia*. The flowers are often spirally arranged on the rachis, even when the leaves are distichous, but the bracts and flowers are distichous in a number of groups, and the flowers are whorled in a few cases, as in *Chamaeangis* and some *Oberonia* species. The inflorescence may arise from any part of the stem. In the primitive condition the inflorescence is terminal and is simply a continuation of the shoot axis (fig. 3.2A, F–K). In other cases the inflorescence is lateral from the side or base of the shoot, or from the rhizome. In the monopodial growth habit, of course, it is always lateral. Sometimes the inflorescence is condensed and the flowers are produced one by one from a very short axis, as in *Epidendrum nocturnum*, *Systeloglossum*, *Bromheadia*, and *Thrixspermum* (fig. 3.12E). The flowers may also be produced simultaneously, or nearly so, in a very dense cluster, as in *Elleanthus* or *Glomera*. Each flower of the genus *Sigmatostalix* arises not from a single bract but from a cluster of bracts, thus probably representing a branch of a more complex inflorescence that has been condensed (F).

The unusual inflorescence found in *Lockhartia* is called cymose, or determinate (fig. 3.12C), a type common among the dicots but infrequent among monocots. The first flower to open is terminal on the inflor-

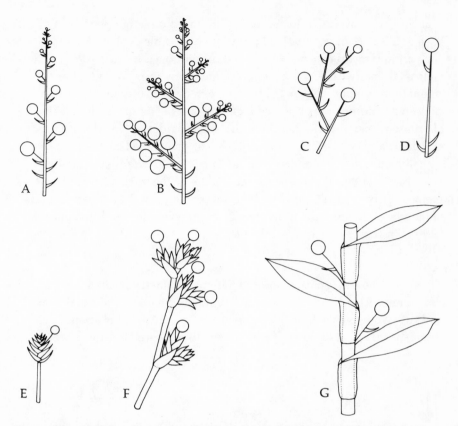

Figure 3.12   Inflorescence types (diagrammatic). (*A*) Raceme. (*B*) Panicle. (*C*) Cymose inflorescence, as in *Lockhartia*. (*D*) One-flowered inflorescence. (*E*) Condensed inflorescence in which the flowers are produced one by one. (*F*) The inflorescence of *Sigmatostalix*, in which each flower arises from a cluster of bracts. (*G*) Leaf-opposed inflorescence, as in *Dichaea*, where each flower arises not from a leaf axil but opposite the leaf axil.

escence, then one or two subtending buds produce shoots which end in flowers, and each of these produces lateral shoots, and so on. Presumably an ancestor of this orchid had a single-flowered inflorescence, and when selection favored an increase in the size of the inflorescence, this was achieved by cymose branching. In the distantly related genus *Notylia* new clusters of flowers are commonly produced from the basal portion of old inflorescences. If the flower cluster of *Notylia* were reduced to a single flower, then the inflorescence would be comparable to that of *Lockhartia*.

While the normal inflorescence is produced from the axil of a leaf or bract, there are some notable exceptions. In *Dichaea* (fig. 3.12*G*) we find that the one-flowered inflorescence is produced directly opposite the base of a leaf (Wirth, 1964). There are two possible explanations for such an inflorescence, and we have, at present, no data that will help us choose. In many cases a leaf-opposed inflorescence represents the ultimate in sympodial growth; that is, each flower is terminal and the next stem internode is produced from the axil of the subtending leaf. The alternate explanation is that the axillary bud "rides up" as the internode elongates and comes to lie opposite the leaf above the leaf to which it is actually axillary. The inflorescence of *Luisia teretifolia* is supraaxillary (Wirth, 1964), but it is borne on the middle of the internode, so that it seems clearly to be derived from the subtending axil. I have seen a similar condition in an unidentified *Epidendrum*. We also find leaf-opposed flowers in some species of *Maxillaria*.

Flowers

The orchid flower typically shows bilateral symmetry; that is, one can draw one, and only one, imaginary line down the middle of the flower and the two halves will be mirror images (fig. 3.13B). This at once distinguishes the orchid flower from flowers like the iris, but there are

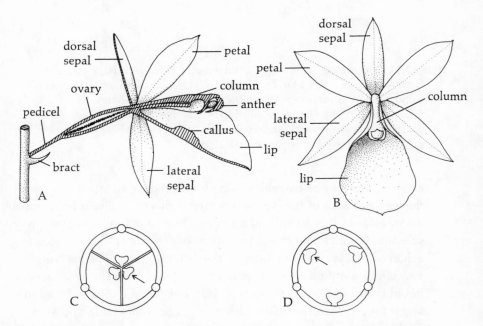

Figure 3.13 Flower parts. *(A)* Longitudinal section. *(B)* Front view of flower. *(C)* Cross-section of a three-locular ovary, showing the placentation. *(D)* Cross-section of a one-locular ovary.

many other flowers, both monocot and dicot, that show bilateral symmetry. In the flowers of *Mormodes*, both the column and the lip are
twisted to one side, thus breaking the normal orchid symmetry. Similar
asymmetry is to be found in the flowers of *Ludisia*, *Macodes*, *Macradenia*, and *Tipularia*, though it is less striking in some cases.

PEDICEL AND OVARY

In the orchids, as in the amaryllids and several other families, the ovary
is inferior—the bases of the other flower parts are completely united
with the ovary so that the other parts appear to arise above the ovary
(fig. 3.13A). Especially in the epiphytes the ovary is not fully differentiated at flowering time, and only continues to develop if the flower is
pollinated. Thus, the plant invests energy only in the ovaries that will
become capsules. For this reason, the ovary is usually slender at flowering time, and it may be difficult to see any distinction between the pedicel and the ovary. An exception is to be found in the subtribe Pleurothallidinae, where there is always a joint, or abscission layer, between
the pedicel and the ovary. In other groups the pedicel is jointed at its
base and falls off with the flower if the flower is not pollinated. The
conduplicate-leaved lady slippers and most Vanilleae, however, have a
joint between the ovary and the rest of the flower; if the flower is pollinated, then the column and perianth fall off after fertilization has
occurred.

As is typical of many monocots, the ovary is made up of three carpels.
In a few primitive genera, such as *Apostasia* and *Selenipedium*, the ovary
is divided into three locules, or chambers, and the placentae are axile
(fig. 3.13C). In most genera the divisions are totally lacking, and the
placentae are parietal (D).

RESUPINATION

We are so accustomed to seeing the lip on the lower side of the orchid
flower that we use the terms dorsal, ventral, and lateral with reference
to this orientation. If the rachis of the inflorescence is erect, the lip is on
the upper (adaxial) side in the young bud, but the pedicel usually twists
as the bud develops, so that the lip comes to be on the lower side. The
term "resupinate" is used for any orchid flower that has the lip on the
lower side (fig. 3.14). The twisting is definitely oriented with respect to
gravity and occurs regardless of the position of the plant or the rachis.
Similarly, most orchids which are nonresupinate also orient their flowers
in relation to gravity so as to maintain their normal posture. In *Catasetum* the pistillate flowers are nonresupinate, while the male flowers are

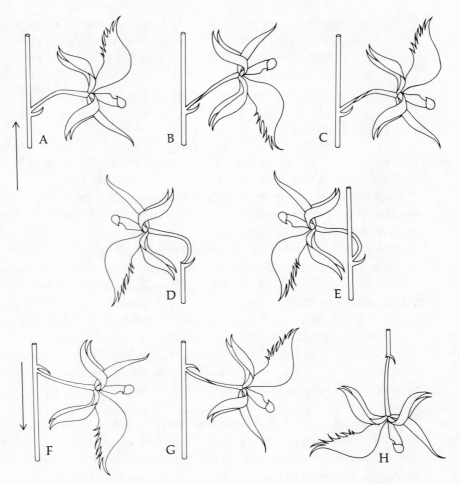

Figure 3.14 Resupination and nonresupination (diagrammatic). *(A,C)* Nonresupinate. *(B,F)* Resupinate. *(C)* Hyper-resupinate. *(D,E)* Resupinate without twisting. *(H)* Pendant, neither resupinate nor nonresupinate. *A* to *E* represent erect inflorescences, while *F* to *H* represent pendant inflorescences.

resupinate in some species and nonresupinate in others. When both pistillate and staminate flowers occur on the same inflorescence, we may observe the pistillate flowers orienting to gravity so as to have the lip uppermost, and the staminate flowers on the same inflorescence doing exactly the opposite. In some cases the pedicel may simply bend down beside the peduncle, so that the flower is resupinate with little or no real twisting. The same sort of bending usually occurs in solitary flowers, such as *Lycaste*, in which case the pedicel bends over the apex of the stem rather than beside it.

Some species of *Malaxis* that have the lip uppermost are actually hyper-resupinate; that is, they have the pedicel twisted through 360 degrees, so that the lip comes to occupy its primordial position (fig. 3.14C). In these species selection has favored a nonresupinate flower, but this has been achieved by exaggerating the normal twisting.

Resupination occurs in all subfamilies of the Orchidaceae; there can be little doubt that this is a basic feature of the family, even though it has been lost or modified in many groups.

PERIANTH

In the Liliiflorae the inner and outer perianth segments are usually similar in color and texture, but when speaking of orchids, one ordinarily uses the term "sepals" for the outer segments and "petals" for the inner segments. In a few primitive genera, notably *Epistephium* and *Lecanorchis*, there is a cupule or calyculus borne at the base of the perianth. A similar structure is found in *Bulbophyllum pachyrhachis* and its allies, but nothing is known of its origin or significance in any of these genera. The sepals have a protective function while the bud is developing, and they are usually valvate, with the edges meeting but not overlapping. In a number of cases, however, they do overlap. The two lateral sepals may be partially or completely united, in which case the term "synsepal" may be used. In other cases, all three sepals may be united or the dorsal sepal may be united with the column. The petals are commonly thinner than the sepals and usually overlap in bud. They may be united with the sepals or with the column, and may be greatly reduced. In *Lepanthes* and *Habenaria* the petals are often deeply lobed and may be much wider than long.

The median petal (opposite the fertile anther) is virtually always differentiated from the other two and is called the lip or labellum. It is usually larger and more complex than the petals and is one of the main elements that makes orchid flowers recognizable as such. The lip is often partially united with the column, rarely with other perianth segments. The lip or part of it may be hinged and movable; in a few cases the movements are active, rather than passive (see Chapter 4). Fleshy lumps, ridges, keels, or plates, usually referred to as the "callus," commonly appear on the lip. Often divided into three or more lobes, the lip is so complex in many orchids that the terms "hypochile" (basal portion), "mesochile" (midportion), and "epichile" (distal portion) have been invented to aid in description. However, the structures so named may not be homologous in different groups.

Darwin tried to explain the complexity of the orchid lip and, at the same time, to account for two of the missing stamens by suggesting that the lip is a compound structure, made up of a petal and two staminodes or sterile stamens. This ingenious hypothesis has had a few proponents, but it has largely fallen by the wayside for lack of evidence. Neither the vascularization nor any other aspect of the lip anatomy indicates that it is a compound structure (Swamy, 1948).

NECTARIES, ELAIOPHORES, AND OSMOPHORES

Since orchid pollen is not used as food by bees, the principal reward offered by orchid flowers is nectar. The lily-like ancestors of the orchids probably had shallow nectar glands between the perianth and the ovary. The most obvious sort of nectary in the orchids is the spur, a slender tubular or sacklike extension from one of the perianth segments, usually the lip (fig. 3.15A). Spurs are well known in *Angraecum* and its allies, *Tipularia, Calanthe*, and many other genera. The members of the Orchideae usually have a single spur formed by the base of the lip, but the related Diseae show remarkable diversity in the form of their nectaries. *Satyrium* has two separate spurs formed from the base of the lip. *Disa* has a spur formed by the dorsal sepal, and *Disperis* has a spur formed by each lateral sepal. In *Comparettia* we find a spur formed by the lateral sepals, which are united, but the nectar is supplied by two slender, solid projections from the base of the lip that extend into the sepaline spur. When we consider pollination and rewards, we will find that a spur is not necessarily a nectary (Chapter 4).

In many Laeliinae the spur-like structure is less obvious, and the nectary runs through the "stem" of the flower, so that it is usually evident only when we split the flower with a razor blade. This type of nectary has been called a cuniculus (fig. 3.15B). In these genera there is a floral tube between the ovary and the base of the perianth. This is most obvious in *Rhyncholaelia digbyana* or *Brassavola cucullata*, which have their flowers borne on long "stems," but the ovary is near the base of the "stem," with a long floral tube between the ovary and the perianth. In these orchids the capsule has a very long beak that corresponds to the floral tube (C). In *Epidendrum* the claw of the lip is normally united with the column to form a tube that is continuous with the cuniculus. In some species the nectary is swollen and externally obvious. A cuniculus type of nectary is also found in the genus *Chloraea* (Diurideae), but in this case there are two parallel, apparently nectariferous tubes, and these extend alongside the ovary, rather than forming a beak between the ovary and the perianth.

In many orchids there is a relatively shallow nectary on the lip or

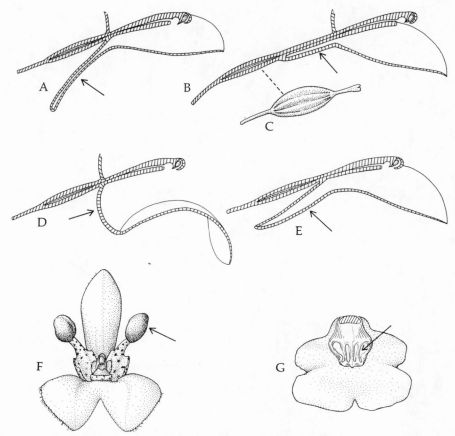

Figure 3.15 Various floral structures. *(A)* Spur formed by the lip. *(B)* Cuniculus. *(C)* Beaked fruit (the beak representing the cuniculus of the flower). *(D)* Column foot. *(E)* Spur formed by column foot and lip. *(F) Pleurothallis raymondii*, showing osmophores. *(G)* Lip of *Sigmatostalix picturatissima*, showing elaiophore.

between the column and the lip. This is typical of *Listera, Stelis,* and many species of *Pleurothallis,* for example. When there is a "column foot," there is often a shallow nectary on that structure, as in many *Dendrobium* and in *Scaphyglottis* (fig. 3.15D). In *Hexisea* and *Systeloglossum* we find deep, spur-like nectaries derived from the column foot or the column foot and the base of the lip *(E)*. In *Dendrobium bigibbum* and its allies we find a spur formed in the middle of the column foot.

Many flower buds have extrafloral nectaries. Field biologists have long suggested that these might attract ants that aid in protecting the flower buds from katydids and other flower eaters. Some antievolutionists dis-

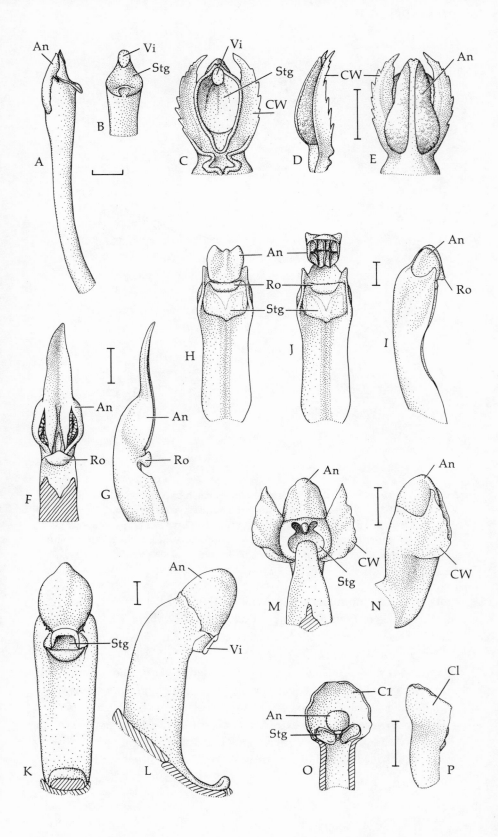

miss this as the wildest sort of neo-Darwinism, but recent workers have obtained experimental data (not with orchids) that support just such an idea. Anyone who has kept orchids under natural conditions will appreciate just how destructive insects can be, and that any extra bit of protection can be helpful. It is quite reasonable, then, to suggest that these glands attract ants, whose presence aids in repelling herbivorous insects.

The shallow, open "nectaries" of *Sigmatostalix, Ornithocephalus,* and some groups of *Oncidium* always have seemed incongruous, since such structures are typical of "promiscuous" pollination systems, not of advanced orchids that are visited by only a few kinds of pollinators. Vogel (1974) has shown that these are not nectaries but "oileries" or elaiophores (fig. 3.15G). As only a few anthophorid bees gather oil, a closed gland is not required to prevent robbery by inappropriate insects.

A third class of floral glands are the osmophores, or scent glands. Perfumes are not produced by all flower parts, or even by the whole perianth, but by specialized areas (Vogel, 1962). The osmophores may be borne on the sepals, the petals, or part of the lip. The characteristic club-shaped tips of the petals or sepals in many Pleurothallidinae are osmophores (fig. 3.15F). When the perfume acts only as advertisement, the exact placement of the osmophore is not too important. In the euglossine-pollinated orchids, where the perfume is both advertisement and reward, the placement of the osmophore is critical, for it determines the placement of the bee in or on the flower. In most cases, the base of the lip produces the perfume.

COLUMN

In all orchids there is some degree of union between the style and the staminal filaments. In most cases these structures are so completely united that we cannot distinguish between them. The combined structure is called the column or gynostemium (figs. 3.13 *A, B,* 3.16). In the Cypripedioideae and Apostasioideae the stamens and staminodes are only partly united with the style. Also, in the Spiranthoideae and in the Diurideae, flowers with relatively short columns have only a very slight

Figure 3.16. Columns of various orchids. *(A,B) Corymborkis,* side view and ventral view of apex. *(C,D,E) Diuris,* ventral, side and dorsal views. *(F,G) Serapias,* ventral and side views. *(H,I,J) Schomburgkia,* side and ventral views with anther in place, and ventral view with anther tipped away. *(K,L) Maxillaria,* ventral and side views. *(M,N) Oncidium,* ventral and side views. *(O,P) Systeloglossum,* ventral and side views. An, anther. Cl, hooded clinandrium. CW, column wing (or staminodium). Ro, rostellum. Stg, stigma. Vi, viscidium. Scale 2 mm.

degree of union. Of *Diuris* itself one can almost say that it does not have a column (fig. 3.16C–E). In most orchids, however, the column is quite obvious, and a good deal thicker than is usually the case with styles or filaments. The column often has lateral wings. These are frequently interpreted as staminodia, and in a number of cases they probably are. In the above-mentioned case of *Diuris*, the wings are virtually free from the column, and it seems very probable that they are indeed sterile stamens. In the case of the Cypripedioideae and *Apostasia*, there is no doubt that the median anther is represented by a staminode.

In most orchids the anther is seated in a clearly defined area at or near the end of the column. This is termed the "clinandrium." In some cases, the edges of the clinandrium are winglike and very prominent; in a few cases they are longer than the column itself, as in some species of *Epidendrum* and *Oerstedella*.

COLUMN FOOT

The base of the column in many orchids forms a ventral extension to which the lip is attached (fig. 3.15D). Sometimes this column foot is longer than the body of the column itself. The bases of the lateral sepals are usually attached to the column foot, and sometimes the bases of the petals as well. When there is a prominent column foot, the flower usually has a spurlike "chin" or mentum when seen from the side (E). The column foot is especially well developed in the *Scaphyglottis* complex and in the Dendrobiinae, where the column foot and the lip together sometimes form a complex structure.

TABULA INFRASTIGMATICA

In the genus *Oncidium* and its allies we find a curious structure known as the *tabula infrastigmatica*, a pad or plate of tissue located ventrally and basally on the column. It is often distinctive in color or texture, and may have a definite function in pollination. In flowers pollinated by anthophorid bees (mostly *Centris*) the bee may seize the base of the column with its mandibles while using its legs to gather oil from the flower. Williams and Dressler (in prep.) suggest that *Oncidium* has evolved from *Trichocentrum*-like ancestors in which the lip was partially united with the column, and that the *tabula infrastigmatica* is part of the lip.

ANTHER

Basically, the anther is an oblong structure with four longitudinal sporogenous sacks, or locules. In the primitive orchids this structure is readily

recognizable as such, but in more advanced groups the anther is modified in various ways. Each anther locule of the Epidendroideae may be
transversely divided into two, so that eight pollen masses are formed,
a condition also found in some species of *Caladenia* and *Eriochilus*
(Diurideae). In this and other groups some anther locules may become
united or suppressed, thus producing fewer pollen masses. Among
primitive orchids the longitudinal partitions which divide the anther into
four cells are evident. In *Coelogyne* and in the vandoid tribes the number
and size of the partitions are reduced, and their orientation within the
anther quite different (fig. 3.17). In the Orchideae, Diseae, and Neottieae
there are small basal appendages, or auricles, at each side of the anther.
These have been interpreted as staminodia, but Vermeulen (1966) notes
that flowers with supernumerary anthers still have auricles on all anthers, and that similar structures occur on the anthers of some Liliaceae.
These auricles include large cells full of needlelike crystals, or raphides,
and may have a protective function. Brieger (in Schlechter, 1970) notes
that the epidendroid and vandoid orchids often have a rooflike extension
of the connective(?) laterally over the anther cells (fig. 3.18). In the
epidendroid group a beaklike "connective wing" is prominent distally
(Hirmer, 1920), and we will see that it is quite functional in pollination.

The attachment of the anther is a feature of some taxonomic importance. In the Spiranthoideae, Neottieae, and Diurideae the anther
commonly is firmly attached to a thick filament, so that it remains in
place after the pollinia are removed (fig. 3.19*A,B*). In the Orchideae and
Diseae the base of the anther is so completely united with the column
that one cannot draw an exact line between column and anther *(C)*. A
similar condition is found in *Acianthus* and *Sunipia*. In most Epidendroideae and Vandoideae, on the other hand, the anther is lightly attached to a very short filament (the rest of the filament being part of the
column) and falls away when the pollinia are removed *(D–F)*. In the
Orchideae the connective is often very broad, and the two pairs of
locules are widely separated.

The position of the anther is traditionally an important feature, though
there is much variation, especially in the more advanced groups, and
developmental studies show the subject to be somewhat complex. In the
primitive condition the anther is erect and parallel to the axis of the
column. This is the case in Spiranthoideae and most Orchidoideae. In the
Diseae the anther may bend over backwards, as in *Disa* and *Satyrium*.
In *Satyrium* the anther actually comes to have the base uppermost and
the apex pointed down toward the base of the flower. In the more primitive Epidendroideae the anther is erect in the early bud, and then bends

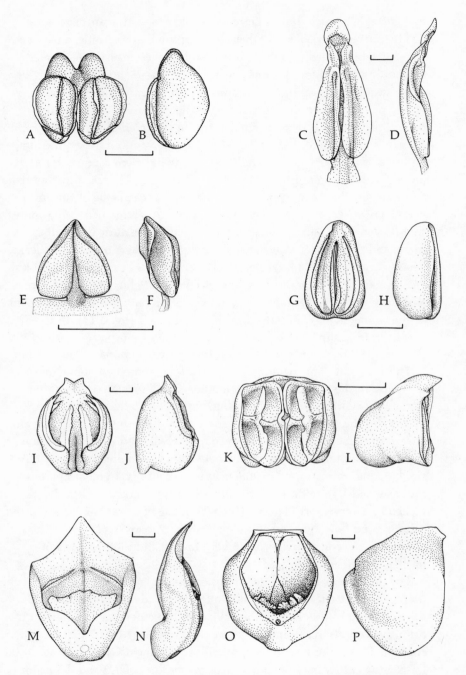

Figure 3.17 Anthers of various orchids. In each case, ventral (clinandrial) and side view are shown. *(A,B) Phragmipedium. (C,D) Sarcoglottis. (E,F) Prescottia. (G,H) Epipactis. (I,J) Sobralia. (K,L) Schomburgkia. (M,N) Cochleanthes. (O,P) Maxillaria.* Scale 1 mm.

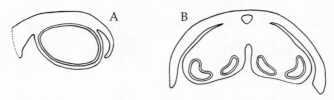

Figure 3.18   Anther of *Vanilla*. (*A*) Longitudinal section through young anther. (*B*) Cross section of anther. (After Hirmer, 1920.)

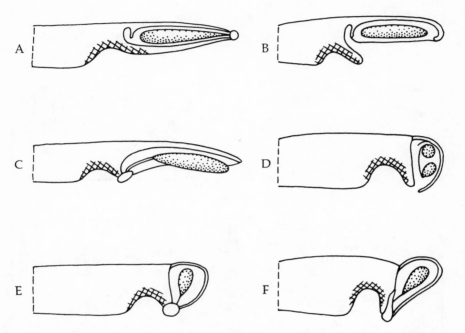

Figure 3.19   Diagrams showing relationships of anther to stigma. Pollen, stippled. Stigma, cross-hatched. (*A*) Spiranthoideae, with anther dorsal and rostellum subequal to anther. (*B*) Neottieae, anther terminal, projecting beyond rostellum. (*C*) Orchideae, with basal viscidia. (*D*) Epidendroideae, with incumbent anther. (*E*) Vandoideae or advanced Epidendroideae, with viscidium. (*F*) Vandoideae, with viscidium and stipe.

downward for 90 to 120 degrees, so that it comes to rest, like a cap, on the apex of the column, or somewhat ventral in its position (*Vanilla, Coelogyne*). The term "incumbent" is used for this condition (fig. 3.20). In the advanced Epidendroideae there are various modifications of the position, and the anther may be erect in many genera, as in *Stelis, Malaxis,* and Podochilinae.

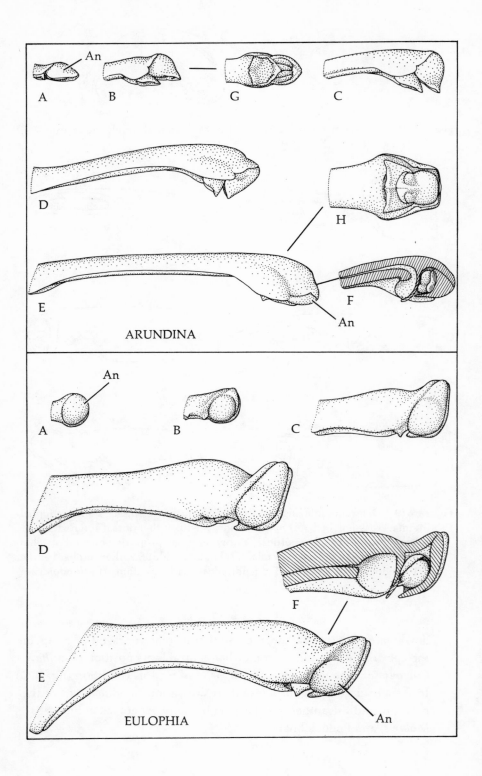

ARUNDINA

EULOPHIA

The term "versatile" is applied to anthers that are attached in the middle and can swivel in various ways. The term has also been used for the anther of the epidendroid orchids. In these orchids the attachment of the filament may be basal, or nearly so, but the anther pivots readily on its point of attachment. When anything brushes the rostellum, it also touches the beaklike connective wing, and this causes the anther to pivot on its point of attachment, bringing the pollinia, or their caudicles, around to touch whatever has just brushed the rostellum (fig. 3.21).

In the Vandoideae the anther usually sits on the apex of the column like a cap, and in every way looks very much like the incumbent anther of the Epidendroideae. Hirmer (1920) has shown, however, that the development is quite different. In most of the Vandoideae one cannot find any trace of the obvious bending that occurs in the Epidendroideae (fig. 3.20). Hirmer interpreted this as the vandoid anther being already bent from its earliest stages. Garay (1972a) states that the anther is bent down, or incumbent, in the Vandoideae but not in the Epidendroideae. In spite of the many resemblances between the anthers of the Epidendroideae and of the Vandoideae, there seems to be a very fundamental difference. The anther of the vandoid orchids does not bend. It remains erect, but dehisces basally rather than ventrally. This is best seen in *Neobenthamia*, where the anther is rather elongate. In many other vandoids the anther has become so short that it looks much like the incumbent anther of the Epidendroideae. Just how to interpret the anthers of some advanced Vandoideae that appear to be erect and dorsal (as in *Notylia* and *Rodriguezia*) I am not sure.

ACROTONY AND BASITONY

It is clear that the base of the anther is next to the stigma in the Orchideae, and equally clear that the apex of the anther is closest to the stigma (rostellum) in the Spiranthoideae and Epidendroideae. Some authors, generalizing from this, have separated the Orchideae and Diseae from all other orchids as the Basitonae (as opposed to the Acrotonae). In the Australian Diurideae, however, we find all degrees between acrotony and basitony (Mansfeld, 1954); the term "mesotony" has been invented for the intermediate condition. Further, the developmental pattern discussed above suggests that the vandoid orchids are quite as basitonic as

Figure 3.20 Development of anther (An) in *Arundina graminifolia* (Epidendroideae) and *Eulophia petersii* (Vandoideae). The anther of *Arundina* is erect in the early stages and bends downward over the apex of the column, while no such bending is shown by *Eulophia*.

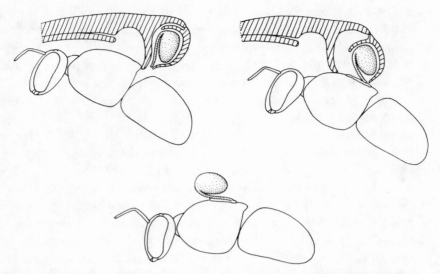

Figure 3.21 The "versatile" anther of the Epidendroideae, which swivels on its point of attachment when an insect brushes against the beak, depositing the pollinia on the glue just removed from the rostellum.

the Orchideae. The relationship between the anther and the rostellum is undoubtedly important, but the family cannot be divided into two major and sharply different categories on the basis of this feature.

POLLINIA

The pollen grains are somewhat sticky or grouped together in some way in almost all orchids, though the term "pollinia" seems scarcely applicable to *Cypripedium* or the Apostasioideae. This aggregation of the pollen grains is adaptively related to the large number of ovules to be fertilized in the orchid ovary. We find the pollen grains as monads, or single grains, in the Apostasioideae, Cypripedioideae, the Vanillinae, some Diurideae, and a few Neottieae and Pogoniinae. In all other orchids the pollen grains either remain in tetrads or are united into larger masses. The structure of the pollen grains has been neglected in the past, but the scanning electron microscope is now being used for this purpose with excellent results (Williams and Broome, 1976; Schill and Pfeiffer, 1977; Schill, 1978; Newton and Williams, 1978). In the Apostasioideae, Orchidoideae, and Spiranthoideae the pollen grains are often heavily sculptured. The Arethuseae also may have more or less heavily sculptured pollen grains, but the Vandoideae and the more advanced members of the Epidendroideae have heavy but relatively smooth exine on the outer walls of the pollinia. The pollen grains of the Vanillinae and the Cypripedioideae are very thin-walled and have relatively little sculpturing. The study of the exine patterns in the more primitive orchids will undoubtedly contribute a good deal of data useful for classification, but it will not solve all our problems. For example, Schill and Pfeiffer (1977), consider the Neottioideae in the sense of Brieger (in Schlechter, 1970) to have rather uniform pollen structure, though I believe that three different phyletic groups are involved here. At the same time, they find the sculpturing to be exceedingly diverse in the Orchideae and Diseae, virtually the only natural group on which all authors are agreed.

We read that the softer orchid pollinia are held together by "viscin" and that viscin is derived from the tapetum, but the term "viscin" has been used in several ways. In most orchid pollinia, and especially in the caudicles, there is a clear, very elastic substance called "viscin" that seems remarkably like rubber in its elasticity. With reference to the Ericaceae and the Onagraceae, the term "viscin thread" is used for strands of sporopollenin which are extensions of the outer wall (exine) of the pollen grain. Balogh (pers. comm.) reports finding sporopollenin strands in the pollinia of the Spiranthinae, but these apparently are not extensions of the tetrad exine. Such strands usually are seen only when the pollinia are broken open, and they are very different from the clear, elastic "viscin" of the caudicles. Ackerman and Williams (in press b) find straplike bands of sporopollenin connecting sister grains within the tetrads of some species of *Chloraea* and *Caladenia*. As "viscin" has been used for different substances, some of them not very viscid, it might be better to abandon the term altogether. Balogh and Nowicke (in prep.) suggest "cohesion strands" for the sporopollenin strands of the Spiranthinae. I shall use the term "elastoviscin" for the clear, elastic material, at least until this mysterious substance can be properly identified.

When the grains are aggregated into distinct pollinia, there are several patterns involved. The relatively soft pollinia may be uniform in structure, as in the Cranichideae, the Vanilleae, and most Diurideae, or they may be "sectile," aggregated into packets, or "massulae," as in the Orchideae, Diseae, Erythrodeae, Gastrodieae, and some Diurideae (*Prasophyllum*; see fig. 3.26). In the massulae and in the harder pollinia of the Epidendroideae and Vandoideae the outer pollen grains have heavy exine on the outer surface of the pollinium, though the sculpturing may be very slight in many hard pollinia. At the same time there may be little or no exine developed on the surfaces between the grains within the massula or in the pollinium (see fig. 3.22). The sectile condition is not intermediate between soft pollinia and hard pollinia; it is a separate

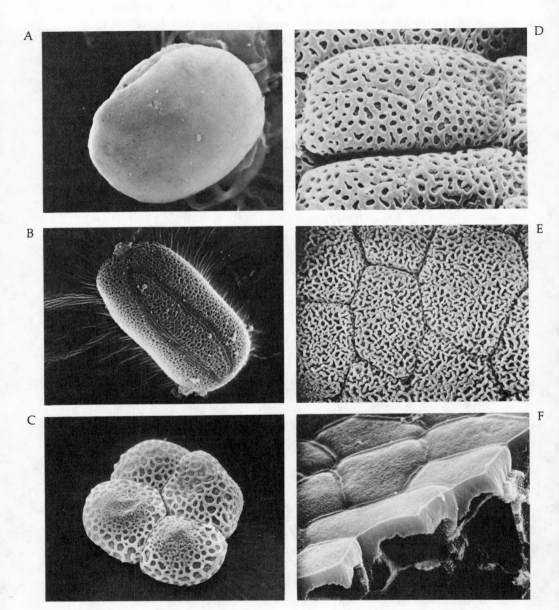

Figure 3.22 Orchid pollen structure: scanning electron microscope photographs. (A) *Selenipedium chica* (Cypripedioideae). (B) *Neuwiedia veratrifolia* (Apostasioideae). (C) *Epipactis microphylla* (Neottieae). (D) *Ponthieva racemosa* (Cranichideae). (E) *Habenaria repens* (Orchideae). (F) *Cochleanthes picta* (Maxillarieae), showing the thick exine on the outer surface of the pollinium, and the lack of a thick exine within the pollinium. (Courtesy of N. H. Williams.)

pattern of adaptation, and a very efficient one. In the case of the sectile pollinia, a single pollinium can pollinate a number of flowers, leaving one or a few massulae in each stigma. The number of ovules fertilized may be less than in the advanced Epidendroideae and Vandoideae, but is still a respectable sum. Sectile pollinia may have a central core of elasto-viscin with wedge-shaped massulae attached at their smaller ends, as in the Orchideae, Diseae, or *Ludisia*. In some Goodyerinae there is a scoop-like "skin" on one side of the pollinium, and the massulae are attached to this skin by a layer of elastoviscin. A similar structure is shown by some Spiranthinae, though the pollinia are not sectile. This scooplike skin, possibly derived from the endothecium (Balogh, pers. comm.), is analogous to the translator of the periplocoid Asclepiadaceae, though by no means homologous.

For the most part, the pollinia are molded by the form of the anther, and the number and shape of the pollinia reflect the form and partitions of the anther itself. In many orchids, there are four pollinia, representing the four anther cells. In many other cases, the four are more or less united into two. In the Epidendreae and the Arethuseae the primitive number is eight, and these may be either laterally flattened or club-shaped (clavate). Within the Epidendreae we find reduction to six, four, or two pollinia occurring independently in the Laeliinae and the Pleuro-thallidinae. There are also many cases in which one finds two or four rudimentary pollinia that are so small as to be virtually nonfunctional. While the pollinia are usually either club-shaped or laterally flattened in the Epidendroideae, in the Vandoideae they are (when four) usually "superposed," that is, flattened parallel to the face of the clinandrium. This same orientation is found in *Coelogyne* and some species of *Sobralia*, and it is not clear just how this orientation has evolved. In the vandoid orchids, fusion of four pollinia into two has occurred, and it is possible that reduction has occurred in the Vandeae also by the loss of one pair, as suggested by Holttum (1959).

CAUDICLES

When the pollinia are relatively hard or compact, there is usually a softer extension or tail by which the pollinia are attached to the viscidium or to insects. These caudicles function both as a "stalk" and as a weak point, permitting the pollinia to break away from a pollinator and stay in the stigma. The caudicles are produced within the anther, and are thus very different from the stipe, which will be discussed below. There is a good deal of structural variation in the caudicles, and the terminology has been a bit confused (see especially Mansfeld, 1934). In a very few cases the caudicles may be hard and bony, like the pollinia themselves. The caudicles may be granular, or made up primarily of pollen grains (which may be abortive), or they may be clear and "hyaline," made up primarily of elastoviscin. In the latter case the caudicles are translucent, very elastic, and lacking in cellular detail. Among Orchideae, at least, the caudicles are partially derived from the wall between adjacent anther cells. Elongate caudicles may be produced from either end of the anther cells. Some Arethuseae and Epidendreae produce caudicles on the ventral side of the anther cell and bear pollinia at each end. Others develop only the basal pollinia (see fig. 9.9), so that the caudicles may be called terminal. Elongate caudicles may be as many as the anther cells (intralocular) or may be produced between adjacent cells (interlocular). In a few cases the caudicles are large and irregular in shape, as in *Coelogyne*. In many other cases they are small and are formed only where the pollinia are attached to the viscidium or stipe, as in most Vandoideae, where the caudicle is nearly hidden in a slit in the pollinium. Caudicles are lacking in the Malaxideae, Dendrobiinae, and Bulbophyllinae, and the pollinia are usually naked. The diversity in structure and form of the caudicles is summarized in table 3.1.

Table 3.1 Different types of caudicles, and some of the genera or groups that show the different combinations of features.

Form	Bony	Granular	Hyaline
Massive		*Coelogyne*	
Point of attachment		*Calanthe* *Pleurothallis*	Vandoideae
Strap	*Neowilliamsia*	*Epidendrum* *Cattleya*	
Cylindrical, intralocular		Goodyerinae	*Fernandezia*(?)
Cylindrical, interlocular		Goodyerinae Orchideae	*Corymborkis* *Cryptarrhena* Orchideae

STIGMA AND ROSTELLUM

The stigma is, as expected, three-lobed, though in many cases the median (dorsal) stigma lobe is much larger than the lateral lobes. One of the most distinctive features of the orchids is the way in which part of the median stigma lobe, the rostellum, has become involved in pollen transfer. There is no rostellum in the Apostasioideae or the Cypripedioideae. In genera such as *Cephalanthera* or *Vanilla* we find no differentiated rostellum, but they show clearly the first steps in the evolution of this structure. In both cases, an insect backing out of the flower will brush the median stigma lobe with its back. This coats the back of the insect with sticky stigmatic fluid, and when the insect touches the pollen, the pollinia or clumps of pollen stick to the insect's back. The next step in this progression we find in genera such as *Sobralia*, *Cattleya* or *Epipactis*. Here the distal part of the stigma projects downward, being different in form and texture from the rest of the stigma. At this point we may speak of a rostellum, noting that it is only a part of the median stigma lobe. In function these flowers are quite comparable to *Vanilla* or *Cephalanthera*. The insect brushes first the rostellum, where it is coated with glue, and then the pollinia or their caudicles. The next step in the series has occurred in every line of evolution of the monandrous orchids, sometimes several times within the same line. A clearly defined part of the rostellum, the viscidium, has the pollinia attached to one side, and its other, sticky side is presented to the pollinator. Thus, the viscidium and the pollinia form a unit and are carried together by the pollinator. Like the rest of the stigma, the viscidium is cellular in origin, but parts of it break down to form the glue. In *Epidendrum* and in many very small flowers the entire viscidium may become semiliquid, and it may then be difficult to recognize the viscidium as such, especially in dried material.

We find an unusual rostellum in the Listerinae, where the rostellar "glue" is held under pressure and is squirted out whenever anything touches the sensitive rostellum. At the same time, the margins of the rostellum reflex, releasing the pollinia, which fall onto the fresh glue droplet (fig. 3.23; Ramsey, 1950; Ackerman and Mesler, 1979). We find another curious device in *Orchis* and *Dactylorhiza*. Here the viscidium is covered by a sheath, the bursicle, that is easily broken away when touched by an insect, thus exposing the fresh, sticky viscidium.

The viscidium in the tribes Orchideae and Diseae is basically in two parts, a condition occasionally found in other groups. This has led Vermeulen (1959) to suggest that the rostellum of the Orchideae is derived not from the median stigma lobe but from the two lateral lobes. However, developmental studies show that the viscidia of the Orchideae are derived from the median stigma lobe (fig. 3.24; and fig. 2 in Vogel, 1959). Vermeulen's drawings of *Coeloglossum* and *Platanthera* show that when the viscidia are far apart the median stigma lobe is extended to them; when the viscidia are together, as in *Dactylorhiza*, there is no corresponding extension of the lateral stigma lobes. Further, the "tape" which connects the viscidia of the Orchideae would be very difficult to explain if the viscidia were derived from the lateral stigma lobes as two

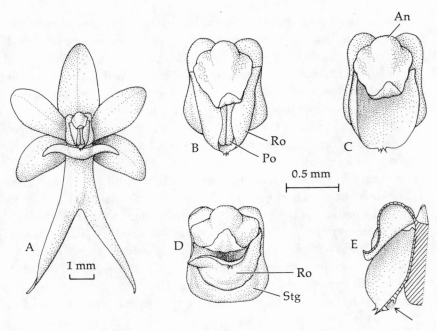

Figure 3.23 *Listera cordata.* *(A)* Flower. *(B)* Virgin column. *(C)* Column immediately after removal of pollinia. *(D)* Column after the rostellum has moved upward and exposed the stigma. *(E)* Longitudinal section of column, showing the gland which is ruptured when the tip of the rostellum is touched. An, anther. Po, pollinia. Ro, rostellum. Stg, stigma. (After Ackerman and Mesler, 1979).

separate structures. Garay (1960) says simply that the Orchideae do not have a rostellum, and that the viscidia are derived from the connective. However, no evidence is offered in support of this curious hypothesis.

The homology of the rostellum was first recognized by Robert Brown. Darwin went further and suggested that the entire third stigma lobe was modified to form the rostellum. Although many recent workers have shown that most orchids have three fertile stigma lobes and that the rostellum is only part of the median stigma lobe (Dressler, 1961; Vermeulen, 1966), Darwin's hypothesis continues to crop up in the literature. In *Habenaria* and *Bonatea*, where two fertile stigmatic surfaces are borne on long stalks, it is possible that these surfaces represent the lateral stigma lobes. In most orchids, however, the median stigma lobe is fertile, and it is much larger than the lateral lobes.

The fertile portion of the stigma usually forms a depression, or in some orchids a flat or slightly convex area. In many Vanilleae, Arethuseae, and Coelogyneae we find what we may term an "emergent" stigma—a membranous structure that projects well away from the body

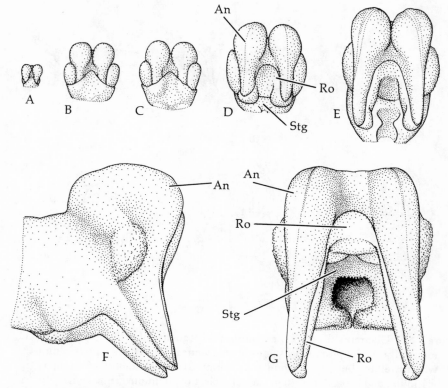

Figure 3.24 Development of column in *Platanthera ciliaris* (Orchideae). The rostellum arises as a single lobe (the median stigma lobe) and elongates to form two separate viscidia. An, anther. Ro, rostellum. Stg, stigma.

of the column. In *Vanilla* the emergent stigma is clearly three-lobed; in other orchids it may be rather an asymmetrical bowl, with the forward edge projecting much more than the rest. The emergent stigma is probably a primitive condition within the Epidendroideae.

STIPE

Elegant as the viscidium may be, it is not the end in our series of clever devices. Most Vandoideae and a few other orchids have a strap of non-sticky tissue attaching the pollinia to the viscidium (fig. 3.25). This is the stipe (the Latin *stipes*, *stipites*, is often used). The stipe is a cellular tissue, derived from the column, and should not be confused with the caudicles. Nearly all vandoid orchids show well-developed stipes, and this feature is often considered diagnostic for the group. A stipe is also found in *Prasophyllum* (Diurideae), and a most interesting parallelism is found in the Bulbophyllinae and *Sunipia*, where one or two stipes are

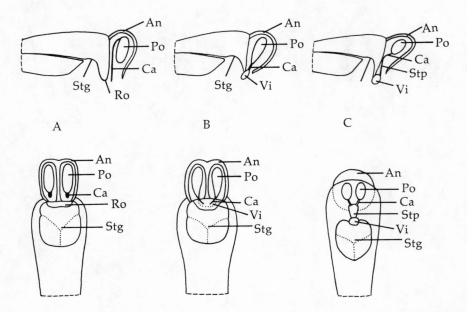

Figure 3.25 Diagrams showing the hypothetical sequence of evolution of the viscidium and stipe; longitudinal sections above, ventral views below. (*A*) Column with rostellum, but no viscidium. (*B*) Column with viscidium. (*C*) Column with viscidium and stipe. An, anther. Ca, caudicle. Po, pollinium. Stg, stigma. Stp, stipe. Vi, viscidium.

to be found in plants which are otherwise very like *Bulbophyllum*. There would appear to be an independent evolution of the stipe in these groups.

POLLINARIA

The actual package carried by the pollinator includes the pollinia, the viscidium derived from the rostellum, and often a stipe as well. The term "pollinarium" has been coined for this unit: the pollinia from a single anther plus all the associated structures that are removed with them. In many Orchideae and in some Vandeae there are two separate viscidia, each with a separate stipe or caudicle and a separate pollinium. Logically, we must introduce a new term here, "hemipollinarium," for in these cases there are two units derived from an anther, and these units may be removed either together or separately. The features of the pollinaria have long been considered of taxonomic importance and, indeed, one can often identify an isolated pollinarium to genus or even to species (figs. 3.26–3.28). With the number and form of the pollinia themselves, and the shape and texture of both the viscidium and the stipe, the pollinarium is a fairly complex structure and includes many different features. Some workers now mount fresh pollinaria on paper triangles, like small insects, so that they may be compared without the distortion that occurs in pressed flowers. In material preserved in liquid, unfortunately, the viscidium and the caudicles often dissolve.

When pollinaria are first withdrawn from a flower, the pollinia are

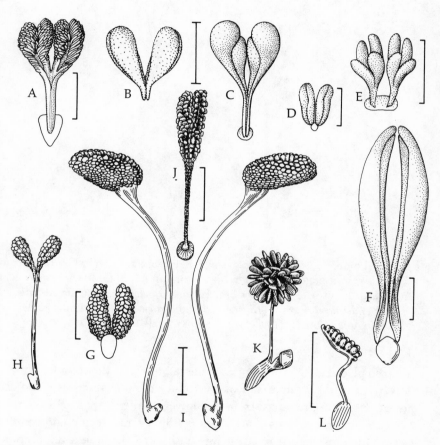

Figure 3.26 Pollinaria of Spiranthoideae and Orchidoideae. *(A) Macodes sanderiana. (B) Prescottia stachyodes. (C) Ponthieva brenesii. (D) Townsonia viridis. (E) Eriochilus cucullatus. (F) Sarcoglottis acaulis. (G) Piperia elongata. (H) Prasophyllum striatum. (I) Habenaria avicula. (J) Disa venosa. (K) Disperis fanniniae. (L) Disperis pusilla.* Scale 1 mm. *(D,E,H after Nicholls, 1969; G after Ackerman, 1977; J,K, after Vogel, 1959; L after Verdcourt, 1968.)*

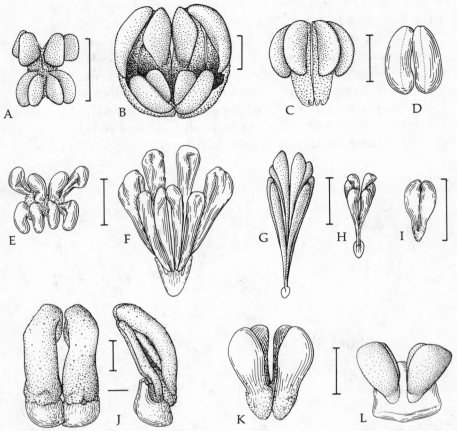

Figure 3.27 Pollinaria of Epidendroideae. *(A) Eria andersonii. (B) Chysis maculata.*
(C) Epidendrum ciliare. (D) Dendrobium fimbriatum. (E) Elleanthus capitatus
(in the broad sense). (F) Calanthe brevicornu. (G) Meiracyllium wendlandii.
(H) Appendicula cornuta. (I) Masdevallia zahlbruckneri. (J) Sobralia powellii.
(K) Coelogyne ochracea. (L) Calypso bulbosa. Scale 1 mm.

often so placed as to strike not the stigma but the anther of another
flower. Thus, the pollinaria of many orchids show characteristic movements after being removed from the flower (Darwin, 1888; Northen,
1970). These movements come about by differential drying and bending
or twisting of either the stipe or the caudicles to change the orientation
of the pollen mass so that it will now strike the stigma of another
flower (fig. 3.29). The fact that these movements require at least several
minutes tends to prevent the insect from pollinating another flower with
the pollinia from that same plant. Perhaps having much the same effect
is the fact that the anther is often removed with the pollinia and does

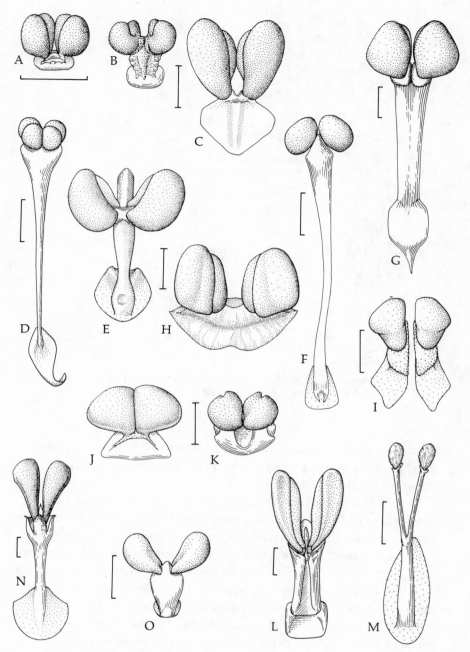

Figure 3.28 Pollinaria of the Vandoideae. *(A) Polystachya bella (B) Govenia liliacea. (C) Warrea costaricensis. (D) Ornithocephalus powellii. (E) Trichoglottis fasciata. (F) Micropera philippinensis. (G) Anguloa dubia. (H) Scuticaria steelii. (I) Dendrophylax fawcettii. (J) Eulophia petersii (K) Grammatophyllum scriptum. (L) Catasetum trulla. (M) Fernandezia sanguinea. (N) Houlletia lowiana. (O) Miltonia regnellii.* Scale 1mm.

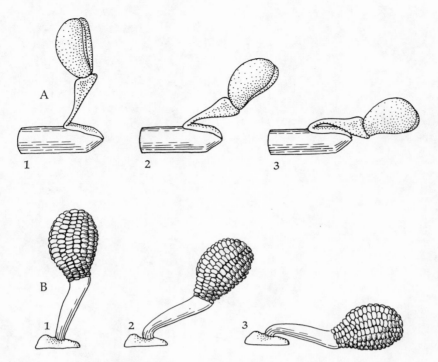

Figure 3.29 Movement in pollinaria. *(A) Himantoglossum hircinum* (after Heusser, 1914). *(B) Rossioglossum grande* (after Northen, 1970).

not fall off until it has dried for several minutes. Also, in a few genera the stipe may coil upon removal from the flower and then straighten slowly, as in *Mormodes*.

The Orchidaceae are not unique in the formation of pollinia. One other plant family, the Asclepiadaceae, or milkweeds, shows an analogous structure. In the milkweeds the pollen grains are held together in a hard, bony matrix, and the retinaculum (or translator) and corpusculum (or "gland") of the milkweeds are formed by a stigmatic secretion which is molded between the stigmatic head and the anthers. Further, the two connected pollinia of an asclepiad are derived from the adjacent halves of two different anthers. Thus, while the two families both have their pollen grains united into pollinia, this union is achieved in quite different ways, and the associated features are very different in their origin and structure. This, and the fact that asclepiad flowers are radially symmetrical, has lead to very different patterns of floral evolution in these two families.

DIMORPHIC FLOWERS

Thre are several cases in which one and the same orchid plant may produce two different kinds of flowers. *Catasetum* and *Cycnoches* are well known for producing male (staminate) and female (pistillate) flowers, often at the same time on the same plant or even the same inflorescence. Intermediate, hermaphroditic flowers are occasionally produced as well. The nature of the flowers is strongly influenced by environmental factors. Large, healthy plants in the sun tend to produce pistillate flowers. Smaller or shaded plants tend to produce staminate flowers (Gregg, 1975). Taxonomists describe the related genus *Mormodes* as having "perfect," or hermaphroditic, flowers, but this genus, too, shows distinct dimorphism in the flowers (Allen, 1959b; Dressler, 1968a). The pistilloid flowers are larger, have a much broader stigma, and do not have hairs on the lip. The staminoid flowers have hairs on the lip in some species and appear to be more short-lived than the pistilloid flowers. Both these types of flower are technically, and perhaps functionally, hermaphroditic, but they would seem to be at a halfway point on the road to unisexual flowers.

In *Oncidium heteranthum, O. abortivum,* and their allies the inflorescence typically includes a few well-developed, functional flowers and a great number of abortive flowers which have only small, straplike perianth parts and no real column (fig. 3.30). These appear to be advertisement flowers which contribute to the splash of color made by the inflorescence but do not represent the energy investment of a fully developed flower. In most groups of *Oncidium* only a few flowers develop capsules, so that the evolution of the abortive flowers would seem adaptive. In *Holothrix burchellii* one finds fertile basal flowers with deeply fimbriate sterile flowers above them (Schelpe, 1966).

In *Grammatophyllum scriptum* we often find a few "abnormal" flowers near the base of the inflorescence, and in *Dimorphorchis lowii* two or three such flowers are regularly found. In the case of *Dimorphorchis* the basal flowers of the pendant inflorescence have shorter and wider petals and a yellow ground color, while the rest of the flowers have a whitish ground color. The yellow basal flowers are strongly perfumed, whereas the other flowers lack any perceptible odor. These flowers retain their perianth and their perfume for a long time, even if pollinated, while the perianth of the other flowers withers soon after pollination. Winkler (1906) suggests that these basal flowers are, in effect, acting as osmophores for the entire inflorescence.

Figure 3.30 *Oncidium heteranthum,* an inflorescense branch with one fully developed flower and several abortive flowers. Costa Rica. (Courtesy of K. S. Walter.)

FRUIT

The structure of orchid fruits has been neglected by botanists. Fruiting specimens without flowers are often difficult to identify to species, but there are several features in the fruit which may be of use in classification. Hallé (1977), in his recent orchid flora of New Caledonia, illustrates fruit structure for most species.

Though the ovary is basically made up of three carpels (and the receptacular tissues surrounding them), this is not at all obvious in the orchid fruit. Rather than splitting between carpels, the orchid capsule splits down the midline of each carpel; and in most orchids the midvein of the carpel, with a little additional tissue, separates from each half-carpel, so that the fruit splits into three wide valves and three narrow ones, though in some cases there are only three wide valves. The fruit usually splits from near the apex and, in some orchids, the edges of the wide valves are connected by transverse fibers through which the seeds must sift. The six valves sometimes separate completely at the apex and spread widely, as in *Lockhartia* and some *Maxillaria* species. In most orchids the valves remain attached apically. Some orchid capsules split along a single line, as in *Angraecum;* in others, as in *Dichaea* and some *Pleurothallis,* the fruit separates into two unequal valves.

In *Cattleya* the wider valves each bear two prominent ridges, so that the fruit is nine-ribbed. In *Encyclia* subgenus *Osmophytum* this same valve usually forms a single high keel, making the fruit triangular in section. Fruits may also be spiny or warty, and the fruits of many orchids are distinctly beaked, the beak sometimes representing a floral tube or cuniculus. Beer (1863) and Malguth (1901) note that terrestrial orchids usually have thin-walled, rather dry and papery fruits, while the epiphytes usually have much thicker, rather fleshy fruit walls. These authors also point out that the fruits of terrestrial orchids are almost always erect, while those of epiphytes are often pendant. In a few terrestrials, as in *Nervilia* and *Corybas*, the peduncle elongates greatly as the capsule nears maturity. This increased height doubtless aids in seed dispersal.

Orchid fruits often have long hairs interspersed among the seeds, termed spring hairs (*Schleuderhaare*) on the theory that they are hygroscopic and that their movements aid in dispersing the orchid seeds. In some fungi the spores are borne in a tangle of hairs known as capilletia. Malguth (1901) compares the hairs in the orchid fruit to fungal capilletia, and uses the term "capilletium" for the orchid hairs. He finds that the capilletial hairs in the fruit of the Vandeae and a few other orchids are strongly hygroscopic, but that the capilletial hairs of most other orchids are not markedly hygroscopic. Malguth finds the capilletial hairs to be characteristic of epiphytic orchids and nearly lacking in terrestrials, even in the terrestrial members of otherwise epiphytic genera.

While 99.9 percent of the orchids have capsular fruits and shed dry seeds, we find a moist pulp associated with the seeds of *Vanilla*, *Selenipedium*, and *Galeola* section *Cyrtosia*; some species of *Neuwiedia* also have fleshy fruits. The presence of fleshy fruits in members of three different primitive groups suggests that the liliaceous group from which the orchids evolved may have had a fleshy fruit rather than a dry capsule. Also we find that very long, slender fruits are often found in relatively primitive orchid groups, as in *Apostasia*, *Selenipedium*, and most Vanillinae.

SEEDS

Seed structure, like fruit structure, has been much neglected. The early work of Beer (1863) gives some idea of the diversity of orchid seeds. More recently, Clifford and Smith (1969), using light microscopy, examined a number of orchid seeds, and Barthlott (1976b) has used the scanning electron microscope to survey a somewhat larger selection. Barthlott concludes that seed structure will be especially useful at the subtribal

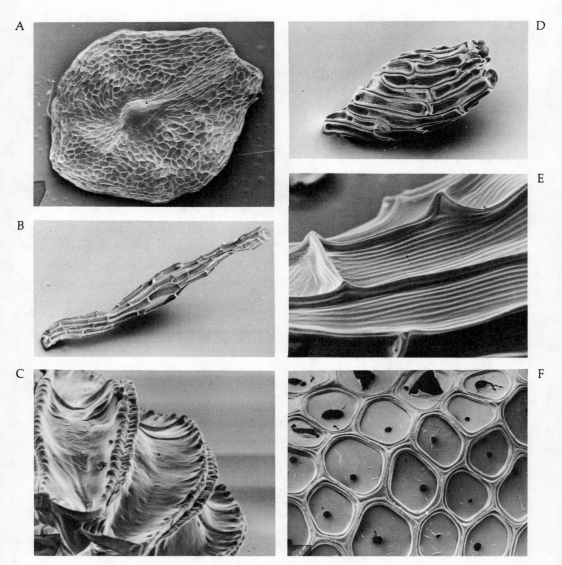

Figure 3.31 Scanning electron microscope photographs of orchid seeds. *(A)*
Epistephium parviflorum (Vanilleae). *(B) Phaius tankervilleae* (Arethuseae).
(C) Disa brevicornis (Diseae). *(D) Mormolyca ringens* (Maxillarieae). *(E) Cyrto-*
podium parviflorum (Cymbidieae). *(F) Coryanthes* species (Cymbidieae).
(Courtesy of W. Barthlott and B. Ziegler.)

and tribal levels. Barthlott and his co-workers have continued with their studies of orchid seed structure, and the publication of their data will be of great interest.

Vanilla, Apostasia, Selenipedium, and some species of *Neuwiedia* have seeds with hard seed coats. The seeds of *Vanilla* are thickly lenticular, and those of *Selenipedium* are angular, both being glossy and dark brown or black. The seeds of *Apostasia* are brown, pitted, and sticky. Other Vanillinae, such as *Epistephium* and *Galeola,* have a hard seed coat over the embryo and a more or less developed wing around the seed. Some species of *Neuwiedia* have small seeds with sacklike append-ages at each end. All other orchids have a loose, rather papery seed coat around the embryo. The seeds range from 0.3 mm to 5 mm in length and vary a great deal in width and in the details of structure. Aside from the size and shape of the seeds, the size and shape of the cells in the seed coat offer features of taxonomic interest, especially in the structure of the cell walls (fig. 3.31). The outer cell wall, for instance, may have longitudinal thickenings, transverse thickenings, or netlike thickenings, or it may be covered with wax deposits, as in most Epidendreae. Barth-lott and Ziegler (in press) report a very interesting structure in the seed of *Chiloschista lunifera* (fig. 3.32). At one end of the seed the seed coat

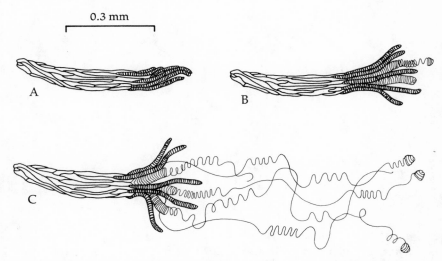

0.3 mm

A

B

C

Figure 3.32   Seed of *Chiloschista lunifera.* (A) Dry. (B) Ten seconds after being moistened. (C) About fifteen minutes after being moistened. The spiral thicken-ings of the specialized "thread cells" become unraveled and form long threads which fix the seed in place once it has fallen on moist bark. (After Barthlott and Ziegler, in press.)

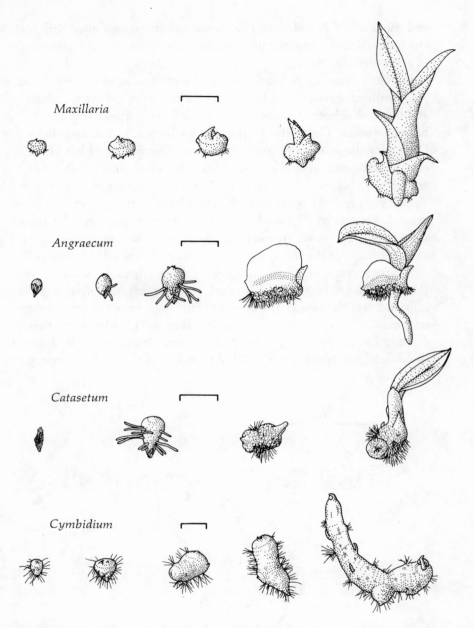

Figure 3.33 Development of seedlings in different orchids. Scale 1 mm. (*Angrae-cum*, *Catasetum*, and *Cymbidium* after Burgeff, 1936; *Maxillaria* after Veyret, 1965.)

cell walls have very strong spiral thickenings. When the seed is moistened, the outer cells spread apart, and the spiral thickenings of the inner
cells become unraveled, forming threads up to four millimeters in length.
These threads doubtless serve to attach the seed to moist bark.

SEEDLINGS

The orchid embryo is not differentiated into distinct organs, as are most
plant embryos. On germination the embryo simply swells into an ovoid
or often top-shaped mass of cells. Root hairs are produced from the
lower portions of this "protocorm." Eventually, a growing point forms
on the upper surface of the protocorm, and a leafy shoot may be formed
(fig. 3.33). Stoutamire (1963) divides the terrestrial seedlings into those
that soon become photosynthetic and those that do not. The quickly
photosynthetic species are characteristic of sunny, wet environments
and may grow rapidly. Most terrestrial orchids, however, do not develop
chlorophyll for several months or more. Stoutamire (1974b) notes that in
many terrestrials with a definite saprophytic stage, the stem formed by
the protocorm bears only scale leaves and grows downward into the substrate. Epiphytic orchids usually become photosynthetic at an early
stage, but some, such as the Catasetinae, have a definite saprophytic
stage in decomposing wood. For the most part, differences in seedling
structure are ecological rather than taxonomic. In the Vandeae, however,
the protocorm is elongate and keeled, and the axis is considered to be
dorsoventral (Burgeff, 1932; Veyret, (1965).

Ecology

4

Ecology, a branch of science concerned with the relationships between an organism and its environment, is an infinitely elastic topic. Because practically anything can be included, discussions of the subject often become too nebulous—and sometimes too philosophical—to be meaningful. Even if we limited ourselves to the definable and measurable, we could easily fill several tomes on the ecology of orchids alone. Here, we will go into the physiological aspects of the orchids' relationship to their environment as-little as possible, except to point out that they, like all other green plants, require light, water, carbon dioxide, and mineral nutrients. We will limit our discussion to those aspects of ecology that have a special relevance to the orchids.

At various points in the discussion of ecology and evolution, I will refer to "problems," "solutions," or "strategies." I do not wish to imply that plants (even orchids) are sentient and plan ways to solve their problems. They are quite passive, but those plants that have, through mutation or recombination, the best strategies, or solutions to their problems are the ones that are most successful in competition and reproduction.

Drainage and Air Movement

While growers find orchids to be tough, adaptable plants, they too often discover that orchids will not tolerate wet, soggy substrates for very long. Under such conditions the roots rot off and the plant declines. Even terrestrial orchids growing in swamps usually have their root systems above the waterline, in a well-aerated medium. Exceptions to the rule that orchids "don't like wet feet" are *Habenaria repens*, *Platanthera flava*, and *Spiranthes odorata*.

Air movement is another critical factor, in both forests and greenhouses, perhaps because of its effect on temperature, gas exchange, or moisture. We do not yet understand the dynamics of this relationship.

Weediness and Stress Tolerance

Botanists often speak of a continuum of plant types ranging from climax species at one extreme to pioneer species (or "weeds") at the other. Among forest trees, climax species are those whose offspring grow up in the shade of their parents and eventually replace them. These species, once established, will maintain the same vegetation type indefinitely unless there is disturbance—a significant destruction of part of the vegetation by such things as rabbits, bulldozers, fires, plows, earthquakes, or hurricanes. Pioneer species, on the other hand, can thrive *only* where there has been disturbance; without it, they eventually disappear, to be replaced by other species. Of course no habitat is totally without disturbance, but this distinction between adaptive types has proved useful, as attested by the amount of recent ecological literature dealing with "r and K selection."

Still, many orchids do not seem to fit this continuum very well. Grime (1977) has proposed a system with three main categories, or "adaptive strategies," that seems better suited for plant ecology. These strategies are: (1) Competitive: Where the habitat is favorable (meaning that water, mineral nutrients, and light are adequate) and there is little disturbance, competition between plants determines which species will occur, and something like the climax vegetation described above will develop. (2) Ruderal (weedy) strategy: Where the habitat is favorable but there is much disturbance, we find weeds, plants that can tolerate and indeed require disturbance. (3) Stress-tolerant strategy: When some environmental factor such as water, mineral nutrients, or light is inadequate, we may say that plants are undergoing stress. Those that can tolerate stress are able to occupy habitats that are unfavorable for the competitive and ruderal plants. There may be competition between plants in a stressful habitat, of course, but adaptations to competition seem less important here than adaptations to other features of the environment.

These are the extreme strategies, and, as Grime points out, most plants fall somewhere between the extremes. Orchids as a group are closer to the stress-tolerant extreme than to either of the others. Most epiphytes, and some terrestrials as well, tolerate a degree of water shortage that would be damaging to many other plants; at the same time, their habitats are often deficient in mineral nutrients. Most of the orchids that do not occupy this type of environment are plants of deep shade, which is quite as stressful a habitat as is a dry, rocky area.

While orchids are rarely the first plants to appear after the forest is cut or burnt, some orchids do show definite weedy tendencies. *Corallor-*

hiza odontorhiza, Liparis liliifolia, and several species of Spiranthes are scarce and very localized in undisturbed habitats but have multiplied greatly in disturbed areas (Sheviak, 1974). Among the tropical orchids, a number occupy steep, rocky sites or open grassy areas; examples are Arundina, Epidendrum calanthum, E. ibaguense, E. radicans, E. secundum, Peristeria elata, Phaius, and Spathoglottis. Under natural conditions, one would expect to find these orchids only in restricted areas. Human activities, however, have produced miles of steep, rocky road cuts and other grassy or rocky areas where these ruderal orchids find a congenial habitat. In the absence of disturbance, they would be much less common than they are. For example, Peristeria elata, the dove orchid, was recorded on Barro Colorado Island (Gatun Lake, Panama) soon after the biological reserve was established and farming was stopped in the area. Now the area is covered with mature or nearly mature forest, and there is no longer any place for the dove orchid.

Mycorrhiza and Germination

The term "mycorrhiza" refers to the symbiotic relationship between a fungus and the roots of a vascular plant. Most plants show this relationship and many, including most forest trees, require it for survival. In the absence of fungi, the seeds of such plants usually germinate and grow well for a few weeks or months but then languish or die unless their roots encounter an appropriate fungus. Most orchids, on the other hand, cannot germinate (in nature) without the fungus, whereas the health of the mature plant seems less dependent on this relationship.

There are several types of mycorrhizae. In the ectomycorrhizal relationship, the fungus is in contact with the root cells but does not penetrate the root, forming rather a sheath of fungal mycelia around the outside of the root. Such mycorrhizae are thought to aid the host plant in the absorption of mineral nutrients from the soil but contribute little or no organic matter to the host. In the case of endomycorrhizae, the fungus does penetrate the host root. The orchid mycorrhizae are of this type, in that the host plant actually digests parts of the fungal body, or else digests cell contents expelled from the fungus (fig. 4.1). Typically, the outer cells of the orchid root form a "fungal host cell layer," which is penetrated by the mycelium of the fungus, but the mycelium in this layer is not digested (Burgeff, 1959). Within that layer one finds the "digestion cell layer," whose cells have been characterized as "phagocytes" by some authors. The mycelium penetrates this layer as well, and then forms dense coils or clumps in which food material (proteins, glycogen, and fat) is stored. These clumps are then digested by the orchid

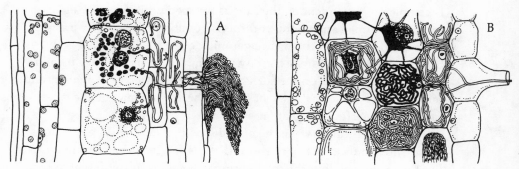

Figure 4.1 Orchid mycorrhizae. *(A) Platanthera chlorantha,* in which the fungus forms clumps that are digested within the orchid cells. *(B) Gastrodia callosa,* in which the fungus expels cell contents into the orchid's digestion cells. (After Burgeff, 1959.)

cell and the foods are used by the orchid—a process that may be repeated many times by the same cell. Beneath the digestion cell layer starch is stored in another layer of cells that is not penetrated by the fungus. In some cases, especially among the tribe Gastrodieae, such clumps are not formed, but the fungus expels part of its cytoplasm (cell contents) into the orchid cells, where it is digested. Clearly, the orchid plant maintains a rather effective control of the fungus and its growth, presumably by means of hormones and other chemical agents. Some authors have gone so far as to characterize the orchids as parasites on fungi, but we do not really know that it is a totally one-sided relationship. The fungi may be receiving vitamins or other substances from the orchid plants.

Because orchid seeds are usually tiny, they cannot contain a large supply of stored foods. When a seed reaches conditions favorable for germination, the first stages of growth may occur in the absence of any fungus, but most orchid seedlings are unable to continue growth without "infection." Some few orchids, especially members of the tribe Arethuseae, are capable of continued growth in the light without mycorrhizae, but their seedlings grow much better with mycorrhizae than without. Typically, fungal hyphae enter the seed through the suspensor end and penetrate the germinating embryo. In some cases the fungus kills the orchid seedling, and in other cases the orchid kills the fungus (and then the orchid dies, unless reinfected). If the proper balance is established, however, the seedling is able to obtain additional food from the fungus and continue growth. For many terrestrial orchids and some epiphytes, the early development takes place within the substrate, and the seedling is completely dependent on the fungus for food until an aerial shoot or leaves are formed and photosynthesis is possible. The seeds of many of these orchids will not commence germination in the light. This is clearly an adaptive feature, for such a tiny seedling would usually dry up and die if it were exposed on the surface of the soil. For many epiphytes and terrestrials of wetter habitats, of course, the surface may be the most favorable place for germination.

The relationship between the orchid seedling and the fungus is not a mycorrhiza in the strictest sense, in that the fungus penetrates not a root but the protocorm. But the relationship is homologous with that in the roots, and it would be silly to invent a new term for it. After roots develop, in most cases they form mycorrhizae with fungi that they encounter in the substrate; the fungus that penetrated the protocorm does not grow out with new roots, though it may be a part of the same mycelium that penetrates the roots. In the tribe Orchideae (and presumably also in the Diseae and Diurideae) the mycorrhiza never occurs in the root-stem tuberoid, and the plant must establish the mycorrhiza anew each growing season when new roots are formed. The tuberoids produce an effective fungistatic agent, especially after other parts of the plant have been "infected" by a fungus (Arditti, 1966a). Similarly, epiphytes have little or no mycorrhiza during the resting season; in *Cattleya* the mycorrhiza is reestablished each growing season (Breddy and Black, 1954). In many cases the mature orchid can grow quite well without any mycorrhiza, at least under favorable conditions of light and nutrient supply.

There has been great controversy over the specificity of orchid-fungus relationships. Some authors, especially Knudson (1922) and Curtis (1939), have maintained that there is no specificity, that the mycorrhiza may be formed with many different fungi, and that the fungus is a parasite and not necessary for the growth of the orchid. It is true that orchid seed sown on agar with adequate mineral nutrients and sugar under aseptic conditions grow quite well. However, such conditions are never found in nature. As to specificity, the situation is not clear-cut. It appears that only a few fungi are compatible with any one orchid species, and related orchid species are likely to be compatible with the same fungi. However, it is very difficult to identify most mycorrhizal fungi. Orchid mycorrhizal fungi have been classified as *Rhizoctonia* species, but *Rhizoctonia* is a "form genus," and when fertile (sexual, sporebearing) material is known, the "Rhizoctonias" are assigned to other genera and families (Warcup, 1975). Warcup (p. 100) sums up the whole

problem of specificity rather well: "Both orchids and fungi differ markedly in the range of partners with which they form effective symbioses."
(See table 4.1.)

Table 4.1 Fungus (Tulasnellales) species known to form mycorrhizae with
orchids. These are the species of which sexual stages have been found. All others
are classified as *Rhizoctonia* species. (After Warcup, 1975.)

Tulasnellaceae	Ceratobasidiaceae	Tremellaceae	[Family?]
Tullasnella allantospora	*Thanatephorus cucumeris*	*Sebacina vermifera*	*Corticium catonii*
T. asymmetrica	*T. orchidicola*		
T. calospora	*T. sterigmaticus*		
T. cruciata	*T. species*		
T. violea	*Ceratobasidium cornigerum*		
T. species	*C. obscurum*		
	C. sphaerosporum		
	C. species		
	Oliveonia pauxilla		

Succulence and Dark Carbon Dioxide Fixation

All green plants depend on photosynthesis for survival. Sunshine and
water are often abundant, leaving carbon dioxide as the limiting factor
in rates of photosynthesis. In many plants the stomata open during day-
light, allowing carbon dioxide to diffuse into the leaves, where it is fixed
and used in photosynthesis. At the same time, of course, the humid
internal atmosphere of the leaf is exposed and water is lost. In the dark,
such plants produce some carbon dioxide by respiration, and most of
this gas diffuses out of the leaf and is lost to the plant.

Many orchids use a different method of carbon dioxide fixation. Because it was first studied in succulents of the family Crassulaceae, this
method is usually known as Crassulacean acid metabolism, or CAM. In
these plants, the stomata are open at night, when the atmospheric hu-
midity is much higher, and carbon dioxide is fixed and stored as malic
acid, making the cell sap strongly acid at night. During the day, the
carbon dioxide is then released and used in photosynthesis. This mechanism reduces water loss and fixes most of the respiratory carbon dioxide
that is produced at night. Further, epiphytes may benefit by fixing carbon dioxide at night, when the forest canopy is producing it. Plants that
show CAM are usually succulent, and always have large photosynthetic
cells with large vacuoles (in which the acid cell sap is stored). In orchids,

the correlation of CAM with thick leaves seems to be very strong: plants with thin leaves do not show CAM, and all thick-leaved species that have been tested do show CAM (Nuerenbergk, 1963; Neales and Hew, 1975). The CAM system is rather flexible; when water is abundant, the stomata may be opened during the day and carbon dioxide enters in the usual manner. In time of drought the stomata of some cacti may be closed day and night, with photosynthesis reduced to recycling the respiratory carbon dioxide. It would appear that succulence and CAM have evolved in many different families, and that these features have evolved independently in different lines of evolution within some families (McWilliams, 1970).

Another efficient system of photosynthesis, the C_4 system is found in sugar cane, maize, and many other tropical plants of sunny habitats, but it is not known in the Orchidaceae.

Epiphytism

An epiphyte is simply any plant that grows upon another plant—usually a shrub or tree—for at least part of its life cycle. This habit of growth is by no means limited to the orchids, though it is one of the best known characteristics of the orchid family. I will not attempt to draw a sharp line between epiphyte and lithophyte, or rock-dweller, since the lithophytic orchids occasionally grow as epiphytes and many epiphytes grow on rocks when conditions are favorable. Epiphytism occurs in 65 different families of vascular plants, involving about 850 genera and nearly 30,000 species, according to Madison (1977). He has estimated the number of epiphytic orchids at 500 genera and 20,000 species. This count may be rather generous, but in any case orchids are known to make up a large part of the world's vascular epiphytes. Other important groups are the Araceae, Bromeliaceae, Cactaceae, Gesneriaceae, the genus *Peperomia* (Piperaceae), and several groups of ferns. Madison finds that the epiphyte flora is less diverse in tropical Africa than in either Asia or America and, somewhat surprisingly, much more diverse in the American tropics than in Asia.

There are several important differences between growing on a tree and growing in the ground, some favorable for plant growth and some not. Light availability is one important difference. In unbroken tropical forest, where little light reaches the forest floor, only shade-tolerant plants can survive except in light gaps (which become areas of strong competition). Because there is constant air movement in the treetops, an epiphyte can tolerate much more sunlight than the same plant at ground level, where fallen plants are often cooked by the sun. Other factors, such as better exposure to pollinators, greater seed dispersal, and avoidance of slugs and other terrestrial herbivores, may also be favorable aspects of the epiphytic habitat (Madison, 1977).

Growth in the tree tops has its unfavorable aspects, too, the most serious of which is probably the lack of water. Except in the wettest (and most constantly wet) cloud forests, epiphytes nearly always have fleshy water-storage organs in roots, stems, or leaves. The epiphytic habitat resembles a desert in some respects, at least intermittently; it is no accident that the orchids often share this habitat with cacti. Benzing (1978) uses the term "extreme epiphyte" to refer to "extremely xeric epiphyte," but it is not clear to me that epiphytes of wetter habitats are any less epiphytic than those of dry habitats. Mineral nutrients are usually in short supply for epiphytes, and most orchids are tolerant of low substrate fertility (Benzing, 1973). Indeed, one finds epiphytic orchids and other epiphytes to be especially abundant in cloud forests and in white-sand areas, both habitats in which terrestrial plants are often stunted by the lack of nutrients.

Some features are common to nearly all epiphytes. The vast majority, including most orchids, are pollinated by animals, and their seeds, which are very small, are dispersed by wind. The adaptive advantage of this last feature is obvious, since large seeds would simply disperse downwards to the soil too often to be functional for full-time epiphytes, whose seeds must lodge on trees or shrubs.

There have been a few attempts to classify epiphytes by ecological category, none of which has seemed successful for the study of orchids. Here we will speak only of three general categories: humus epiphytes, which grow only where there is a soil layer on the tree bark; bark epiphytes, which occur on trunks and large branches without much humus; and twig epiphytes, which are found on very small branches and are, of necessity, micro-orchids. These are not sharply defined categories, but they are sometimes useful in discussing the growth of epiphytes. The humus epiphytes are, in a sense, the least specialized, and with them one often finds other plants that are not normally epiphytic.

A few Malaysian orchids are restricted to single host-tree species (Went, 1940), though this is rather the exception in America and Africa (Allen, 1959a; Johansson, 1975). In Brazil some species of *Pseudolaelia* and *Constantia* seem to be restricted to stems of *Vellozia* (Velloziaceae), just as *Cymbidiella pardalina* is normally restricted to the staghorn fern *Platycerium madagascariense* in Madagascar, but both of these hosts have unusual physical properties. In the American tropics the smaller species of *Psygmorchis* seem to be restricted to the twigs of *Psidium guajava,* but these tiny plants are easily overlooked, and they would be very hard to find in the forest canopy, if they occur there. All this does not mean that orchids are usually distributed at random on all tree species. Some trees are good orchid hosts and regularly carry a heavy load of epiphytes, while other species in the same areas bear few or none. An orchid species that is largely restricted to one host tree in a given area may "prefer" another host tree under different climatic conditions (Sanford, 1974). Orchids often thrive when transplanted onto tree species that are not usually host trees (Allen, 1959a). Presumably, germination is the most critical stage, and mature plants may grow well on trees that are not suitable for germination. Bark and twig epiphytes are most strongly correlated with the host-tree species; the humus epiphytes require only a layer of humus, and it matters little what tree species is beneath the humus (Went, 1940).

The physical properties of the bark are undoubtedly important. A soft, spongy bark with roughened surface is best for water retention and also provides cracks and crevices where orchid seeds may lodge. Smooth bark is less favorable, especially in drier habitats, and trees that frequently shed their outer layer of bark give orchids little chance to become established. The spongy barks of *Acnistus* (güititi), *Crescentia* (calabash), and *Paragonia* (Bignoniaceae) are all especially favorable for orchids in Central America. In Thailand, *Elaeocarpus grandiflorus* (Elaeocarpaceae) is an exceptionally good orchid host.

Chemical factors are less obvious but are at least as important as physical properties. Orange trees, for example, are good orchid hosts, while the physically similar lemon trees rarely bear orchids. In a Mexican cloud forest *Quercus castanea* and *Q. vicentensis* have many orchids, *Q. scytophylla* and *Q. peduncularis* have fewer, and *Q. magnoliaefolia* bears no orchids, bromeliads, or mosses and very few lichens (Frei, 1973a; 1973b). The bark structure of these oak species is similar, but the bark of *Q. magnoliaefolia* contains gallic and ellagic acids which inhibit orchid germination. *Quercus peduncularis* and *Q. scytophylla* contain fewer inhibitor substances than *Q. magnoliaefolia,* but the bark of *Q. scytophylla,* at least, has some inhibitory effect on orchid seedlings.

As one moves from drier to wetter habitats, the physical structure of the bark is probably less critical, and even moderate amounts of inhibitors may be leached out, while the presence of other epiphytes becomes much more important. Many Mexican orchids grow not directly on bark or rock but upon lichens (Pollard, 1973). Thus, the inhibitory qualities of some oak bark may act primarily on lichens, whose presence is necessary for orchid germination. It is probably the physical features of the

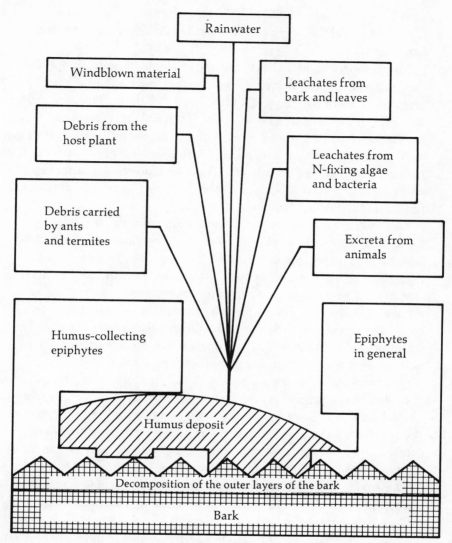

Figure 4.2 Origin of substrates and nutrients of importance for the epiphytic flora. (After Johansson, 1975.)

lichens, especially their water-holding capacity, that promote orchid germination, but no detailed studies are available. In wetter habitats the bark is commonly covered by mosses, liverworts, and lichens, so that the epiphytic orchids are not necessarily in direct contact with the bark. In the extremely favorable habitat of cool cloud forests, the distinction between epiphyte and terrestrial breaks down, as "terrestrials" grow upon the lower tree trunks and "epiphytes" thrive as well on the mossy ground as in the trees. Similarly, the best place to collect "epiphytic" orchids at moderately high elevations in the Andes is often on steep road cuts.

Large, horizontal branches may have an abundance of epiphytes on their sides or hanging from the lower surfaces, while their upper surfaces bear only moss. Perry (1978) points out that monkeys, squirrels, and other arboreal mammals regularly destroy plants growing on top of the branches in order to keep their pathways open, especially if the tree or its near neighbors produce edible fruits or seeds. Perry also notes that hollow trees inhabited by bats often have an unusually luxuriant epiphyte load, presumably because the bats are supplying rich, nitrogenous fertilizer to the whole canopy.

Physical conditions may vary greatly even within a single host tree, and careful study will usually show distinct zonation (Johansson, 1975; figs. 4.3, 4.4). A few orchids, such as *Cochleanthes, Pescatorea,* and *Aspasia,* thrive in the weaker light and greater moisture of the lower tree trunk. Some species, especially the more massive plants such as *Acineta, Cymbidium,* and *Grammatophyllum,* are usually found only on very large branches or in crotches. Other species are more abundant on branches of moderate size, and many micro-orchids occur primarily on twigs just within the leaf canopy. These patterns are largely controlled by physical factors such as light and moisture, which may vary a good deal over short distances. In eastern Chiapas, Mexico, for example, several *Lepanthes* species were found on twigs of freshly fallen trees in the tall tropical evergreen forest. But in an open pine and hardwood forest nearby, these same *Lepanthes* species occurred only two to four meters above the ground on the twigs of *Hauya heydeana* (Dressler, 1957). And in the same area, many of these micro-orchids which were widespread in the treetops could also be found in a localized "elfin forest" on limestone cliffs overlooking a lake.

Only a few years ago northern biologists seemed to have a very mystical concept of the tropical forest, viewing it as a vast, uniform, unchanging greenhouse where some supernatural agency carefully cultivated all the trees and avoided any nasty competition. Of course that is not the case. Tropical forests are real habitats, with real plants growing in them. Aside from extreme cold, the tropical forests are exceedingly diverse in every feature. Tropical soils are by no means uniform; local rock outcrops or white-sand areas produce functionally dry pockets in the wettest forest, and nearly all tropical forests have a dry season of some sort, sometimes two in each year, which critically affects epiphytes. The most

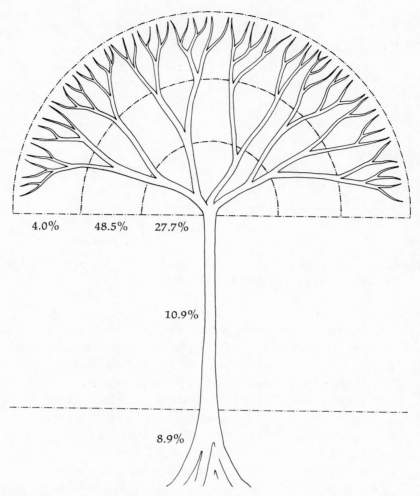

Figure 4.3 Different zones in the host tree, and the percentage of orchids found in each zone in a West African forest. (After Johansson, 1975.)

outstanding single feature of the tropical forests is not their uniformity but their extreme biotic diversity.

Epiphytosis

Throughout tropical America orchids and other epiphytes are commonly known as parasites. The botanist often objects, saying that they do not harm the tree. Of course, orchids and other epiphytes may harm their host trees by providing too much shade, by keeping the branches too

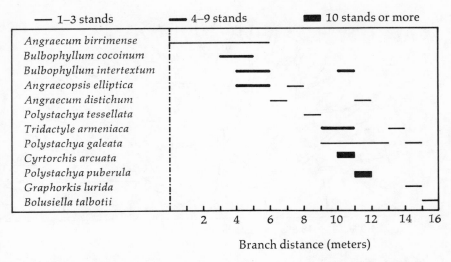

Figure 4.4 The number of orchids per meter along a 16-meter-long branch of *Parinari excelsa,* in West Africa. (After Johansson, 1975.)

wet, and by overloading the tree, causing breakage. But even this damage may not be the whole story. Ruinen (1953) has found convincing, if largely circumstantial, evidence that some orchids and other epiphytes are harmful to tree crops. Ruinen suggests that the mycorrhizal fungus penetrates the host tissues and that the orchid is indirectly acting as a parasite. Johansson (1977) has suggested that *Microcoelia exilis* is harmful to its host trees in West Africa. Perhaps the epiphyte inhibitors found in some trees have a very real adaptive value.

The orchids most often accused of harming their host trees are the twig epiphytes, such as *Microcoelia, Ionopsis,* and *Taeniophyllum,* precisely those orchids that seem, at first glance, to be least attached to the host tree. Clearly, the twigs and smaller branches offer the fungus much better opportunities for penetration than larger branches with thick layers of bark. It is interesting that the leafless Vandeae are accused of hemiparasitism both in Africa and in Asia. Johansson (1977) suggests that the loss of leaves may have evolved in conjunction with this indirect parasitism. These leafless Vandeae can be cultivated on nonliving substrates, so there is no doubt that they can photosynthesize and survive without a living host under favorable conditions.

Ruinen (1953) also suggests that epiphytes may compete with their host trees for limiting mineral resources. Benzing and Seemann (1978) have coined the term "nutritional piracy," and suggest that such competition may be critical on poor soils. They find that some *Tillandsia* species manage to get enough of the available nutrients to severely affect the growth of the host trees. This is doubly convenient for the epiphyte: by getting the mineral nutrients that it needs, it also stunts the growth of its host tree, thereby getting more sunlight and prolonging its own life. While Benzing and Seemann discount any allelopathic effects of the epiphytes on their host tree, the production of inhibitory chemicals could be quite adaptive for the epiphytes. The more sun-loving epiphytes, especially, could prolong their life spans appreciably if they could slow their host's growth rate and postpone heavy shading by the host's foliage.

Saprophytism

The term "saprophyte" is used for any plant that does not manufacture its own food by photosynthesis but uses organic material (previously manufactured by other plants) in its substrate. Thus, most fungi are saprophytes, at least if one considers them to be plants. The term is also commonly used for those flowering plants that lack chlorophyll, such as snow plant and Indian pipe (Monotropaceae). In fact, no vascular plant is capable of digesting and absorbing the organic material in the leaf litter, and some authors therefore prefer to use a different term, such as "mycotrophic," for these nonphotosynthetic vascular plants. Such plants always have mycorrhizae and use, as food, substances that are digested and transported by the fungi. Orchids normally have a saprophytic stage in their seedling development, and in some orchids this stage may be quite long. It is not surprising, then, that wholly saprophytic plants occur in various orchid groups. In such plants, the saprophytic stage is prolonged throughout the life cycle, and in most cases the plant remains completely concealed within the substrate except at flowering time. In many of these saprophytes the roots are absent or poorly developed, and the much-branched stem has a coral-like appearance. Some of the mycorrhizal fungi are able to digest lignin and cellulose, and some saprophytic orchids, such as *Galeola*, are commonly associated with logs or dead trees. Similarly, some of the fungi associated with these saprophytes are known to be parasitic on living trees, and the orchids may well be receiving much of their nutrition indirectly from tree roots. Some of the better known saprophytic orchid genera are *Corallorhiza, Galeola, Gastrodia, Hexalectris,* and *Neottia.* There are also a number of genera, such as *Cephalanthera, Cymbidium,* and *Eulophia,* which are largely autotrophic (photosynthetic) but include one or a few wholly saprophytic members in each genus. Also, there have been several reports of albinistic individuals of *Cephalanthera* and *Epipactis* that lack chlorophyll but are nevertheless able to survive and flower, even though the plants are smaller than their green neighbors.

The greatest concentration and variety of saprophytic orchids are to be found in tropical Asia and Australia, where the relatively gigantic *Galeola* is found; most other saprophytes are small and inconspicuous. There are few saprophytic orchids in Africa. The systematic distribution of saprophytism within the orchid family is also interesting. No saprophytes are known in the subfamilies Apostasioideae or Cypripedioideae, and only a few are known in the Spiranthoideae. In the Orchidoideae, the Neottieae include several saprophytes, but only a few are to be found in the Diurideae and Orchideae. Among the Epidendroideae, the Gastrodieae are nearly all saprophytic, as are the Epipogieae, and there are many saprophytes in the Vanilleae, but only a few in the Arethuseae *(Hexalectris)*. The remaining Epidendroideae lack saprophytes, and among the Vandoideae the more primitive subtribes of the Maxillarieae (Corallorhizinae) and Cymbidieae (Cyrtopodiinae) each include several saprophytes. In these groups and in the subtribe Catasetinae, the saprophytic stage is quite prominent, even in the autophytic genera. Thus, *Eulophia, Govenia, Oecoclades, Catasetum,* and allied genera form quite large coralloid masses in the soil or within rotting wood before producing an aerial shoot, which, when it finally appears, may be quite large. These orchids sometimes grow to flowering size in a surprisingly short time, perhaps because of the initial saprophytic stage.

The saprophytic orchids cause special problems for the taxonomist. The leaves are quite rudimentary, and stem and root structures are greatly altered, so that vegetative anatomy is of very limited utility in determining their relationships. In many cases they are self-pollinating, making it difficult to determine the details of column structure. And because the plants are inconspicuous and virtually impossible to cultivate, little material is available for study. In older classifications they were often grouped together without regard for the details of flower structure.

Ant Plants and Trash Basket Plants

Evolution has given rise to several interesting solutions to the problem of severely limited mineral nutrients in some tropical habitats, especially among the epiphytes. One such adaptation is ant symbiosis. While some other epiphytes in especially mineral-poor environments exhibit a more striking relationship with ants than do orchids (see Janzen, 1974), several orchids do show a definite association with ants. Docters van Leeuwen (1929) notes that *Acriopsis javanica* is a regular inhabitant of the nests of *Crematogaster*, and *Dendrochilum pallidiflavens* is said always to grow with ants of the genus *Iridomyrmex*. In both cases the seeds have oil droplets that make them attractive to the ants, so that they carry the seeds to the nest. Ridley (1910) observes that ants tend to build their nests in the matrices provided by the roots of orchids, such as *Dendrobium crumenatum*. He considers the plants associated with ants to be healthier and to suffer less in periods of drought than their less fortunate neighbors. Janzen (1974) observes that some orchids, such as *D. crumenatum*, may "parasitize" more specialized ant plants, such as *Hydnophytum* or *Dischidia*, when growing near them, by sending their roots into the debris dumps of the ant plants.

The American tropics are characterized by "ant gardens," where *Codonanthe* (Gesneriaceae) and species of *Anthurium*, *Epiphyllum*, *Aechmea*, and *Peperomia* flourish (Ule, 1904; Wheeler, 1921; Weber, 1943; Kleinfeldt, 1978). Some of these plants are rarely found except on ant nests, and most of them have seeds with elaiosomes, or oil bodies, which are dispersed by ants. The ant nests provide a favorable habitat for germination, and the resulting root mass reinforces the ant nest, which often attains the size of a football or even larger (fig. 4.5). The ant gardens are commonly formed on nests of *Campanotus* and *Crematogaster* (less frequently on nests of *Azteca*, *Anachetus*, and *Solenopsis*). Interestingly enough, ants of these two genera are often found in the same nest, with each ant species using different but interconnecting cavities within the nest and often sharing foraging trails. In most cases the larger *Campanotus* occupies the core of the nest, while *Crematogaster* lives nearer the surface. A number of orchid species may be found in such ant gardens, but *Epidendrum imatophyllum* and species of *Coryanthes* are virtually never found anywhere else and are difficult to cultivate in the absence of ants. Not only do these species benefit from the increased mineral supply brought to the nest by the ants, but the ants also protect the roots and other parts of the plant from other insects. These plants can be cultivated without ants if extra fertilizer is added, an acid medium is maintained, and insecticides are regularly used (Dodson, pers. comm.). Other orchids such as *Epidendrum baumannianum* and *E. schomburgkii* and species of *Sievekingia* often occur on ant nests and seem to reach their best development there, though they are not limited to this habitat. In the American tropics, too, epiphyte roots make a good matrix for other types of ant nest. A moderately large ponerine ant, *Odontomachus*, is often met by orchid collectors, who usually leave the associated orchids in peace.

A few orchids show special physical adaptations to ant symbiosis.

Figure 4.5 *Coryanthes speciosa* shares an ant garden with *Peperomia macrostachya* and a small *Epidendrum imatophyllum* in Central Panama.

Both *Schomburgkia* section *Chaunoschomburgkia* and *Caularthron* (*Diacrium*) have hollow, tapering pseudobulbs with openings near the base where the ants may enter (fig. 4.6). These plants are usually inhabited by ants and surely benefit from both improved mineral nutrition and some degree of protection, but they grow quite well in the absence of ants.

Another way to beat the mineral shortage is for the plant to form a "trash basket" and directly collect debris, which then forms humus about the plant roots, where the minerals released by decay are readily available to the plant. None of the orchids is as efficient a trash basket as are many ferns, anthuriums, and bromeliads, but a few do show

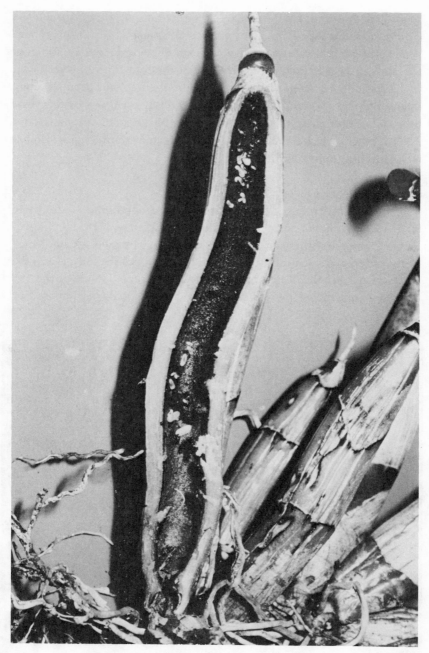

Figure 4.6 A pseudobulb of *Caularthron bilamellatum,* split open to show the ant colony within.

special adaptations to trash collecting. *Grammatophyllum, Ansellia, Graphorkis, Cyrtopodium,* some species of *Catasetum,* and some members of the subtribe Stanhopeinae, for example, produce thin, stiff roots that grow upward and outward from the substrate and serve very well as debris collectors (fig. 4.7). In other cases the stems or pseudobulbs themselves may accumulate some trash, but few form an obvious "birds nest." Many ferns are rather efficient trash-basket plants. Johansson (1974) indicates that some of these ferns are important in supplying sites where orchids may germinate and grow.

Phenology of Flowering

One of the most frequent questions heard by an orchidist in the tropics is "When do orchids flower?" The answer, of course, is that orchids bloom throughout the year, but each species has its own season. Much of the older, and even rather recent, literature gives the impression that there are no seasons in the tropics. This is not true. Outside of the tropics, as one moves toward either pole, the seasonality is imposed pri-

Figure 4.7 Aerial roots of *Grammatophyllum papuanum,* which form a sort of trash basket that collects debris. Lae, Papua New Guinea.

marily by cold. Within the tropics, on the other hand, seasonality is primarily due to variation in rainfall. Truly aseasonal climate is extremely localized. There are a number of ecological reasons (apart from the physiological reasons, which do not greatly concern us here; see Rotor, 1952; Arditti, 1966b) why each species should have a more or less definite season. A few trees and vines can afford to produce a few flowers every day or every few days for many months, but this is exceptional, even with very large plants. Plants that do not flower constantly need to coordinate the different members of a population in some way; cross-pollination is not possible if the different individuals of a species all flower at different times. As a matter of fact, flowering at different seasons is an important mechanism by which closely related species are able to occupy the same area without forming hybrids and losing their identity. In Costa Rica, *Cattleya skinneri*, which flowers in February and March, and *C. patinii*, which flowers in September and October, would seem to be a closely related pair that overlap geographically but do not form hybrids (Fowlie, 1967).

Some seasons may be much more favorable for flowering or for the dispersal and germination of seeds (an often overlooked aspect of flowering phenology) than others. In nontropical areas, winter imposes a rigid seasonality on plant growth, and virtually all flowering occurs during the growing season. Stoutamire (1974b) notes that most nontropical orchids release their seeds at the beginning of the dormant season, in autumn. Flowering, however, occurs at different times during the growing season, depending on the species. In many tropical areas, the dry season is not so drastic as to impose absolute dormancy. In West Africa, Sanford (1971) finds that most orchid species flower during the rainy season, especially during the early and mid rainy season. Johansson (1974) finds a similar pattern, with two flowering peaks in the rainy season. Dunsterville and Dunsterville (1967), who recorded the flowering of 280 orchid species for about two years of observations, conclude that relatively few orchids have a short, once-a-year flowering period and that most flower over a period of two to several months each year, whenever they feel like it throughout the year (but never in time for the orchid show). The Dunstervilles' data, however, are based on cultivated plants, many of them outside their normal geographic range and all protected from normal stress by tender, loving care. The only data I can find on tropical American orchids under natural conditions are given by Braga (1978; see fig. 4.8). Here again we find some species recorded in flower during only one month, and one during six months. There are fewer orchids in flower during the dry season than during the rainy

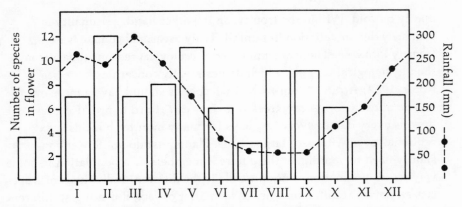

Figure 4.8 Number of orchid species in flower each month in a "campina" near Manaus, Brazil. (After Braga, 1978.) The average monthly rainfall in Manaus is indicated by the dashed line.

season, but the difference is not very great (nor is the dry season that severe near Manaus). Das (1976), who recorded the flowering of *Coelogyne* species at Shillong over a number of years, reports that different species are in flower throughout the year, with a number of species commencing flowering in March (spring), and that some species are in flower for less than a month, while others flower for as much as six months (table 4.2).

It is my impression that in Panama the later rainy season (October and November) is the most unfavorable season of the year. Too much rain and too little activity by pollinating insects produce a definite lull in the flowering and fruiting of most plants, including orchids. Many Central American orchid species flower at the end of the rainy season or the beginning of the dry season (December and January). In wetter habitats, such as cloud forests, the dry season may be much the most favorable for flowering, having less rain, better visibility for pollinators, more active pollinators, and fewer plant-eating insects (which show a predilection for buds and flowers). The dry season may also be a favorable time for seed release, as there is usually more wind then. In drier areas, only epiphytes with well-developed storage organs can afford to flower during the dry season—in the drier parts of Panama, rat-tail Oncidiums *(Lophiaris)* and *Encyclia cordigera*, for example.

Changes in day length are often the cue for flowering. A number of tropical orchids are known to be short-day (actually long-night) plants, but other factors, such as temperature, may also serve to stimulate flowering. In truly aseasonal areas such as Singapore it is much more

Table 4.2. Flowering seasons of *Coelogyne* species at Shillong, India, by fortnight. (After Das, 1976.)

Species	I	II	III	IV	V	VI	VII	VIII	IX	X	XI	XII
graminifolia	+	+	+									
micrantha		+	+	+								
flaccida		+	+	+								
cristata		+	+	+								
elata			+	+	+	+						
ochracea			+	+	+	+						
punctulata			+	+	+	+	+					
corymbosa			+	+	+	+	+					
prolifera				+	+	+	+					
longipes				+	+	+						
suaveolens					+	+	+					
flavida						+	+					
occulata						+	+					
rigida						+	+	+	+			
ovalis							+	+	+	+	+	+
fimbriata									+	+	+	+
barbata										+	+	+
fuscescens	+											

difficult to synchronize flowering, but even there the weather is not abso-
lutely uniform. Many species flower in response to dry spells or to cool,
wet periods and thus achieve synchronous flowering, at least over large
areas. There are, of course, many different flowering patterns in the
orchids. For those species that flower for many weeks or have very long-
lasting flowers, a degree of synchrony is not difficult to achieve. In many
euglossine-pollinated orchids of tropical America, the flowering season is
not at all sharply defined and the inflorescences are usually quite short-
lived, but they are highly attractive to pollinators. Such a system seems
especially well suited to promote cross-pollination over considerable
distances (Dressler, 1968b; Williams and Dodson, 1972). Any given
inflorescence may fail to be pollinated (or to supply the pollinia for
pollination), but several inflorescences are commonly produced during
the ill-defined season.

The ultimate in synchrony is achieved by the orchids that show gregarious flowering. These plants produce short-lived flowers but are syn-
chronized to the point that all or nearly all plants of a species flower on
the same morning. In the American tropics this pattern is shown by
Sobralia and *Triphora*. In Malaysia it is shown by *Bromheadia*, *Diplocaulobium*, *Ephemerantha*, *Thrixspermum*, and some species of *Dendrobium* (Smith, 1925; Coster, 1926; Holttum, 1949). One of the best
known and most studied cases is that of the pigeon orchid, *Dendrobium
crumenatum*. Its flower buds develop to a certain stage and then cease
growing until the right cue is presented, a period of cool weather accompanying a rain storm or a rainy period. Then the flower buds resume
development and, since all start at the same stage, all flower on the same
day, nine days after the drop in temperature. A number of different
species may use the same cue without flowering on the same day, if they
have different periods of development after the cue (see table 4.3). This
seems to be the case also in *Sobralia*. There are often a number of sympatric species, but each species seems to have its own day for flowering.
Dodson has noted that the occasional plant that misses its day and
flowers by itself seems to be ignored by potential pollinators. On
Sobralia day, though, pollinating insects fix on the abundant flowers, and
many are visited and pollinated. Gregarious flowering is not entirely
limited to orchids, but the only nonorchid that I know to show such
exact synchrony is *Napeanthus* (Gesneriaceae).

Flowering in Response to Fire

Fire caused by natural events such as lightning is very much a part of
the environment in many areas, especially in "Mediterranean" climates,
with their dry summers and rainy winters. There, one finds some plant

Table 4.3 Different species of gregarious-flowering orchids, and the number of days required for their buds to open after a drop in temperature. (After Coster, 1926.)

8 days	9 days	10 days	11 days
		DENDROBIUM	
acuminatissimum	crumenatum	spurium	carnosum
papilioniferum	fugax		
spathilingue	insigne		
		DIPLOCAULOBIUM	
brevicaule	aratriferum	dendrocolla	
compressicolle	filiforme	dilatatocolle	
crenulatum	nitidicolle	ecolle	
validicolle			
		EPHEMERANTHA	
bicostata	kelsallii	forcipata	angulata
comata			xantholeuca
fimbriata			
flabelloides			
luxurians			
macraei			
maculosa			
		THRIXSPERMUM	
arachnites		raciborskii	
calceolus		subulatum	
inquinatum			

species that flower or disperse seeds only after fires. Such species, of course, would become extinct if fire were completely eliminated from their areas. Schelpe (1970) notes that in South Africa the terrestrial orchids show three patterns of reaction to fire: (1) The plants flower in season irrespective of fires. (2) The plants flower more prolifically after a fire, but continue to flower in subsequent years without fires. (3) Some species apparently flower only after a fire. Fire undoubtedly releases mineral nutrients, and it may also clear the land, so that herbaceous plants receive more sun. Either effect could stimulate increased flowering in some orchids. Both South Africa and Australia have a number of orchids which flower more abundantly in response to fire (Schelpe, 1970; Stoutamire, 1974b), and in Africa, at least, some species seem to be strictly dependent on fires for flowering and reproduction.

Pollination
A critical step in the evolution of life on earth was the development of sexual reproduction. By permitting the recombination of genetic material

from different individuals, sexual reproduction makes possible the genetic variation on which natural selection must act. Variation occurs in nonsexual organisms, to be sure, but without recombination, evolution is glacially slow compared to the evolution of sexually reproducing organisms. The problems of sexual reproduction are far different for plants than for motile animals. Animals can move about and seek a mate, whereas plants must entrust their pollen to wind, water, or some moving organism. This presents an acute problem in tropical forests, where the great diversity of plants means that neighboring plants are likely to be of quite different species.

Orchids are not normally pollinated by wind or water; all (except self-pollinating plants) are pollinated by animals, especially insects. Many of the most striking features of the family are adaptations to animal pollination (van der Pijl and Dodson, 1966). Darwin devoted much time and energy to studying the pollination of orchids and this doubtlessly played an important role in the formation of his concepts of evolution.

Advertisement and reward are two important factors in animal pollination. We may discuss them in anthropomorphic terms if we bear in mind that no real thought or planning is involved on the part of the plant, and very little on the part of the animal. The pollinating animal must first notice and visit the flower, and so the flower is advertised by its color, form, and fragrance. However, animals will not visit a particular kind of flower repeatedly unless they can expect to receive a reward of some sort. Normally the reward is food, usually nectar or pollen, and animals will quickly learn to avoid a kind of flower that never has any reward.

FIDELITY AND RESTRICTION

We may use the general term "fidelity" for any tendency on the part of a pollinator to return to the same kind of flower. "Flower-constancy" refers to a sort of fixation, which may be temporary or long lasting, and especially to food gathering from relatively abundant flowers by social bees. While such behavior may be quite efficient for some bees and plants, it is probably not very significant in orchid pollination. Some degree of fidelity, however, is critical.

From the insect's point of view, a guaranteed reward would be ideal, and from the orchid's point of view, repeated visits by the "best" pollinator for that species would be ideal. Consequently, in many cases orchid flowers are so constructed that only the "right" pollinator—small flies, medium-sized butterflies, or large bees, for example—can obtain the reward. This feature of flower construction is called "restriction," and it is advantageous to both parties. It has been said that most orchid species have their own specific pollinators, but this is an exaggeration. Some of the most specialized orchid species are each pollinated by a single insect species, but because none of the insects is strictly limited to the pollination of a single plant species, the specificity even in these cases is one-sided.

Since different kinds of animals are shaped differently, see different colors, are attracted to different smells and shapes, and so on, one can often make an educated guess as to the pollinator of a flower from the flower's features. Thus, pollination ecologists speak of "syndromes." Some people have scoffed at the syndromes because of insects that visit the "wrong" flowers and because of flowers that do not fit well into a single syndrome. Still, the syndromes are quite useful, if one remembers that all sorts of animals, whether arthropod, avian, or primate, often stick their noses into something from sheer curiosity; such behavior is often adaptive for the animal. There is no reason why a hummingbird should not occasionally visit a "butterfly" flower or a butterfly probe a "hummingbird" flower. Indeed, these two types of pollinator are really very similar (from the flower's viewpoint), and often overlap in the types of flowers that they visit. If a long-tongued bee wishes to sup nectar from what is primarily a butterfly flower, it will do so. And if bees consistently visit these flowers more often than butterflies, it is likely that natural selection will shift the features of the plant population toward the bee-flower syndrome. Similarly, hummingbirds are aggressive and inquisitive animals, occasionally probing in all sorts of flowers. In general, though, when bees visit the flowers that are well-suited to bees, it is advantageous to both pollinator and pollinatee. Such syndromes, as they apply especially to the Orchidaceae, are summed up in table 4.4.

As a whole, orchids have an elegantly simple pollination system. If a certain behavior results in removal of pollinia from a flower, the same behavior in another flower of the same species will result in the deposition of pollinia in the stigma. In the "gullet" flowers of many orchids—those that form a kind of chamber into which the pollinator must enter—there is an added refinement: the bee does not receive or deposit pollinia when it enters the flower but only on its withdrawal. Thus, self-pollination (of the same flower, at least) is highly unlikely (see fig. 3.21).

Aside from "bee" flowers, "bird" flowers, "fly" flowers, and so on, floral ecologists often classify flowers by form and function, independently of the kind of pollinator involved. These categories include "flag" flowers, "brush" flowers, and "gullet" flowers. Here we will be concerned only with a couple of these categories (fig. 4.9). The term "gullet" flower applies to such flowers as snapdragon, penstemon, and many other members of the dicot orders Scrophulariales and Lamiales. In most

Table 4.4 Some aspects of the syndromes of features associated with different pollinator classes, with special reference to orchids.

Characteristics	Bees (feeding)	Euglossine males	Butterflies	Moths	Nectar flies	Carrion flies	Birds
Odors	Sweet, day	Resinous, morning	Agreeable, day	Sweet, strong, night	Sweet or disagreeable, day	Foul, day	None
Color	Diverse, including ultraviolet, no pure red	Diverse	Vivid, red, yellow	White, cream, pale green	Green, yellow, brown, red-purple	Dull, brown–purple	Vivid, red, yellow, cerise
Form	Gullet	Diverse	Tubular	Tubular	Cupped, shallow	Cupped or trap	Tubular or narrow cup
Landing platform	+	+/–	+/–	–	+/–	+/–	–
Nectar	+	–	+	+, abundant	+	–	+, abundant
Nectar concealed	+/–		+	+	–	–	+
Nectar guides	+	–	±	–	–	–	–
Other	May have pseudo-pollen				Fringes	Fringes, slots, or windows	

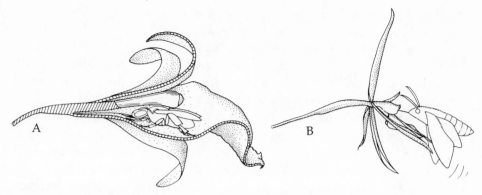

Figure 4.9 *(A)* A gullet flower, in which the column and perianth form a chamber which the pollinator must enter. *(B)* A key-hole flower, in which the placement of the pollinator is controlled by the small opening that gives access to the reward.

cases, the pollen is deposited on the back of the insect, but it may also be placed on the head or the ventral surface in some flowers. *Cattleya* and *Sobralia* represent a sort of gullet flower in which the chamber is formed by the lip and the column alone. The other perianth parts function only as advertisement (at least after the bud opens). Another category, the "keyhole" flower, involves a narrow opening and more or less tubular structure which will admit only the tongue, beak, or proboscis of the pollinator. Thus, the pollinator is rather precisely positioned in or on the flower, and the pollen is usually placed on or near the face of the polli-nator, or on the proboscis or beak itself. Many bird flowers, moth flowers, and butterfly flowers fall into this category—as do *Epidendrum*, the African angraecoids, and many Orchideae. Floral ecologists some-times use the terms "nototribic" for flowers that place the pollen on the dorsal surface of the pollinator, and "sternotribic" for those that place it ventrally on the pollinator. These logophiles could have a field day with orchids. The compaction of the orchid pollen into discrete bodies with sticky pads built on makes the pollen placement of orchids much more precise than that of most other flowers; with the euglossine bees alone, one can easily distinguish thirteen different modes of pollinarium placement (fig. 4.12).

POLLINATORS

Now we may briefly review the cast of pollinators that are most impor-tant in orchid pollination:

Hymenoptera. Bees, which are essentially pollen-gathering wasps, are the flower visitors *par excellence*. The females visit nonorchid flowers industriously to gather pollen to be used as food for their larvae, and both males and females gather nectar, either for themselves or for their larvae. Orchid pollen is virtually never used as food by flower-visiting insects, so that other rewards or attractants, including pseudopollen and imitation pollen, are involved in all orchid-bee relationships.

The symmetry and form of orchid flowers suggest that the first steps which set this group apart from the other Lilialean monocots were adaptations to pollination by bees or wasps. Even now, about 60 percent of the family are normally pollinated by bees or wasps (van der Pijl and Dodson, 1966). Bee-pollinated orchids are often gullet flowers or modifications of this type. There are many sizes and types of bees, from tiny halictines and stingless bees up to massive carpenter bees, large *Centris*, and the larger euglossine bees. Of the estimated 19,000 to 20,000 species of bees in the world, a few groups deserve special mention. Carpenter bees, of the genus *Xylocopa*, are very widespread and are the usual pollinators of a number of orchids, especially large gullet flowers such as *Arundina, Cymbidium, Eulophia, Schomburgkia*, and *Vanda*. The familiar bumblebees, *Bombus*, are often pollinators of *Spiranthes* and related genera, and especially of northern and high-elevation orchids. Two subgroups of the family Anthophoridae, including the genera *Centris* and *Paratetrapedia*, gather oils from flowers instead of (or in addition to) nectar. Several American orchids have evolved for pollination by these bees. These include *Ornithocephalus, Sigmatostalix*, and some groups of *Oncidium* that have open "nectaries" (actually elaiophores) on the lateral lobes of the lip (Vogel, 1974).

Euglossine bees. The tribe Euglossini, of the family Apidae, appear to be more closely allied to the bumblebees than to any other group. This group of five genera and about 180 species is limited to tropical America. They range from small (housefly size) to very large. They are mostly long-tongued, fast-flying insects, and they have left their mark on the floral evolution of many tropical American plant families. The females pollinate some nectar-producing orchids, such as *Sobralia*, but they are far less important in orchid pollination than the males. The males gather perfume droplets from the surfaces of flowers (and from other sources, when they can find them) and store these liquids in their hollow hind legs. These substances are undoubtedly used by the bees in some way, probably as pheromones or sexual odors, after some modification (Dodson et al., 1969).

In the American tropics, the members of the subtribes Catasetinae and Stanhopeinae are all adapted to pollination by euglossine bees, as are most members of the Zygopetalinae and some Oncidiinae. Instead of nectar, all these orchids produce a strong, often resinous perfume, especially during the morning hours. These perfumes are powerful attractants for euglossine males, who brush on the flower surface with their front feet, hover in the air while transferring the perfume to the hind legs, and then land and repeat the activity several to many times. This has permitted the evolution of bizarre and complicated systems of pollination such as those of *Gongora, Stanhopea,* and *Coryanthes,* as well as the more conventional systems of *Acineta* and most Zygopetalinae. The perfumes of these orchids are blended in such a way that each orchid species attracts only one or a few species of pollinator. With as many as fifty species of euglossine bees in a single area, and with the physical diversity of pollination systems (fig. 4.10), there is ample opportunity for orchid speciation.

Analysis of orchid perfume has permitted the identification of many of the chemical components that attract euglossine males (Hills, Williams, and Dodson, 1968; Holman and Heimermann, 1973). Several of these substances, when used by themselves (without blending with others), are powerful attractants for the bees, and many seemingly rare bee species have been collected by the hundreds in recent years. This permits us to study orchid pollination without any orchids. By collecting those bees that bear orchid pollinaria and identifying their pollinaria, we can determine which euglossine-pollinated orchids are flowering at any given time and which bees are visiting them (Dressler, 1976a; fig. 4.11).

Diptera. The flies, or Diptera, include a very diverse group of insects. The mosquitoes are, as far as the flower is concerned, rather like tiny moths, and in the far north they may substitute for moths, as in the pollination of *Platanthera obtusata* (Thien and Utech, 1970). Other flies are specialized flower visitors, such as the Syrphidae, Bombyliidae, and some Tachinidae. These look and act much like bees, and pollinate in much the same ways. Many small flies or gnats are associated with the fruiting bodies of fungi. These flies may seek nectar in flowers, but they are strongly attracted to fungi, and normally lay their eggs on fungi, which are food for the larvae. Small, nectar-seeking flies (whether fungus flies or not) pollinate the great majority of the Pleurothallidinae and many Bulbophyllinae. Most pleurothallids are pollinated by tiny, *Drosophila*-like flies, and we still know very little about their specificity or behavior.

The other class of flies which concern us here are the carrion flies. These do not normally visit flowers for nectar, but some flowers have evolved a way of using them as pollinators. These flowers mimic rotten

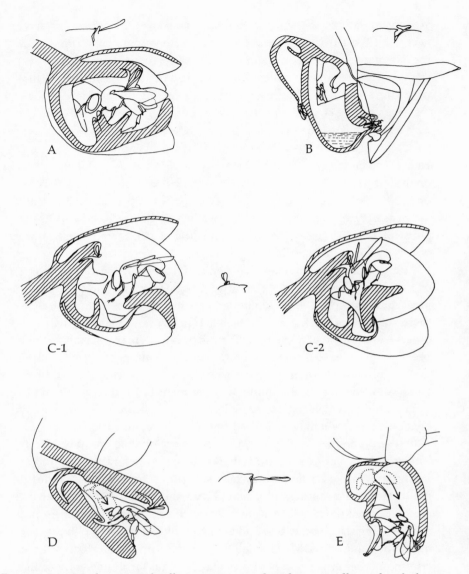

Figure 4.10 Mechanisms of pollination in several euglossine-pollinated orchids. In each case, the placement of the pollinarium on the bee is shown in an inset. (A) *Acineta*. (B) *Coryanthes*. (C) *Peristeria*. (D) *Stanhopea ecornuta*. (E) *Stanhopea oculata*. (After Dressler, 1968a.)

flesh, in odor and often in color, and thus attract flies. Many *Bulbophyllum* and quite a few species of *Masdevallia* fall in this category. Their flowers are either dull or lurid and their perfume can best be described as nauseating. Unpleasant odors (unpleasant at least to humans) are

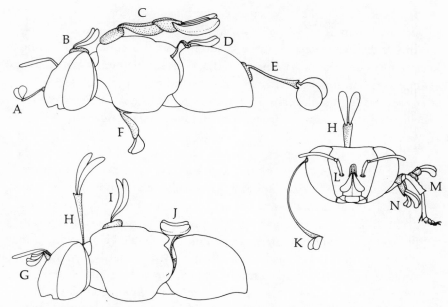

Figure 4.11 Pollinaria on euglossine bees, showing different modes of placement. *(A) Notylia. (B) Peristeria elata. (C) Catasetum. (D) Acineta. (E) Cycnoches. (F) Dressleria. (G) Kefersteinia. (H) Kegeliella. (I) Peristeria. (J) Coryanthes. (K) Macradenia. (L) Trichocentrum. (M) Clowesia warczewitzii. (N) Chaubardiella.*

sometimes associated with nectar flowers as well, such as *Listera*. Presumably, such odors are considered both pleasant and attractive by the pollinators.

Lepidoptera. The butterflies and many moths regularly visit flowers to obtain nectar, and are nearly as well known as flower visitors as the bees. These insects have a long tongue, or proboscis, which is coiled into a tight spiral when not in use, and flowers adapted to them usually have a narrow spur and are of the keyhole-flower organization. Butterflies usually carry pollen on their tongues, and in other plant groups butterfly pollination is normally associated with flowers that contain few ovules, as the butterfly tongue is structurally unsuited to carry large masses of loose pollen. Orchids avoid this problem by the formation of pollinia. Butterflies, and a few moths as well, are diurnal insects and, unlike the bees, they can detect red as a color. Thus, butterfly flowers are often pink, red, or yellow. Butterfly flowers may have perfumes, but visual advertisement seems to be more important in most cases. The majority of moths, on the other hand, are nocturnal and, at least at night, can only distinguish light from dark. Moth flowers are usually white or pale green, and tend to be strongly perfumed at night. The sphinx moths (family Sphingidae) are very strong-flying moths, and some authors draw a distinction between the syndrome of sphinx-moth pollination and that of pollination by ordinary moths. Both sphinx moths and skippers may carry ordinary pollen on their hairy bodies, but the bodies are poor sites for the deposition of sticky orchid viscidia. This may, incidentally, be the main reason that orchids have not adapted to bat pollination.

In the American tropics, *Brassavola* and many *Epidendrum* species are usually pollinated by moths, as are many of the Angraecinae and Aerangidinae in Africa. Something very like the relationship between orchids and male euglossine bees occurs with some groups of *Epidendrum*. In *Epidendrum anceps* and in the *E. paniculatum* complex the flowers are regularly visited and pollinated by male Lepidoptera. *Epidendrum anceps* (in Florida, at least) is visited primarily by males of a single species of Ctenuchid moth, *Lymire edwardsii* (Adams and Goss, 1976). The members of the *Epidendrum paniculatum* complex attract male Ithomiid butterflies, but the attraction appears to be less specific than in *E. anceps*.

Birds. In both the Old World tropics and in tropical America there are groups of birds that specialize in flower visiting, and, in turn, flowers that are specialized for bird pollination. The syndrome for bird pollination is similar to that for butterfly pollination, but the flowers tend to be much stiffer, and perfume is usually lacking. Bird pollination is more important at higher elevations in both hemispheres. In Malaysia and New Guinea a number of *Dendrobium* species show the typical bird-pollination syndrome and are surely visited by sunbirds (family Nectariniidae). The New World hummingbirds (family Trochilidae) are highly specialized flower visitors, and a number of American orchids are adapted to hummingbird pollination. These include *Elleanthus*, *Cochlioda*, *Comparettia*, and some species of *Epidendrum*.

In bird pollination the pollinia are normally deposited on the bird's beak, and among hummingbird-pollinated orchids there is a high percentage of species with dark, slate-colored, or greyish pollinia, suggesting that natural selection has favored pollinia which match the color of the bird's beak, bright yellow pollinia presumably being wiped off more frequently by the bird (Dressler, 1971). Interestingly enough, the *Dendrobium* species of highland New Guinea which show other features of the bird-pollination syndrome, such as *D. phlox* and *D. sophronites*, also have dark, slaty pollinia.

ORCHID HABITATS VERSUS POLLINATOR HABITATS

Cool, wet cloud forests seem to be the optimum orchid habitat; at least that is where the orchids show greatest abundance and maximum diversity. Bees, on the other hand, are most at home in dry, sunny areas. Therefore birds become more important in pollination at higher elevations. Cruden (1972), comparing bird-pollinated and bee-pollinated flowers at high elevations in Mexico, found that in good weather both classes are well pollinated; but when flight conditions are bad, as they often are at high elevations, the bee-pollinated flowers are very poorly visited. Hummingbirds, being endothermic, or warm-blooded, are relatively independent of weather and remain active even under cloudy or rainy conditions. In Central America euglossine bees are abundant and diverse up to perhaps 1100 meters in elevation, but above that their number and diversity diminishes rapidly; they are rarely seen above about 1500 meters. This distribution is reflected in the occurrence of euglossine-pollinated orchids, as well as other euglossine-pollinated groups, such as the Marantaceae. In marginal areas—cloud forests from 1000 to 1200 meters in elevation, for example—the bees may be active on a sunny day, but on the many cold, wet, cloudy days they are either inactive or, more likely, foraging elsewhere.

The flowers of the genus *Stanhopea* usually last about a day and a half, but the flowers of some populations of *S. anfracta* found at relatively high elevations in Ecuador last for four or five days, thus much increasing their chances of being in flower on a sunny day (Dodson, pers. comm.). Whereas the perfumes of euglossine-pollinated orchids at lower elevations tend to be chemically complex and rather specific, attracting only one or a few species of pollinator, some of the orchids of cloud forests have perfumes that are strong and much less specific, as in *Houlletia odoratissima, Coeliopsis,* and a *Peristeria* species that occurs in Panamanian cloud forests (Dressler, 1968a; 1968b). The perfume of the *Peristeria* smells almost like pure 1–8 cineole, and has attracted as many as sixteen species of euglossine bee in a short period of observation. In the earlier analysis, I concluded that orchid species which are sympatric with other species of the same genus have more specific perfumes, suggesting that natural selection has favored greater specificity among orchids with close sympatric relatives. However, many of the orchids which do not have close allies growing with them are the species of cloud forests, and this undoubtedly skewed the results of my analysis. It seems rather that natural selection has favored a perfume as attractive as possible in the marginal habitats, where flight conditions are often unfavorable. At much higher elevations, of course, the euglossine bees, and most other bees, are absent. In these habitats, some orchids are pollinated by bumblebees, but fly pollination and bird pollination seem to be the predominant modes at very high elevations. Flies are limited by weather conditions, also, but they are everywhere. Self-pollination occurs in some high-elevation species, but it seems to be infrequent.

NONPOLLINATORS AND ANTIPOLLINATORS

Early observations tended to list all insects seen on or near a given flower as pollinators, but floral ecologists have learned now to draw a sharp distinction between visitor and pollinator. This distinction is relatively easy to make when observing orchids, because the pollinia are large enough to be seen on the pollinator (or one can detect their absence from the flower immediately after visitation). In the euglossine-pollinated orchids, it is not unusual for a nonpollinator that is much too large or too small to effect pollination to be attracted to a flower and brush on the surface of the flower. Nonpollinators that are too small probably have very little effect on fertility, though visitors that are too large may sometimes frighten away the legitimate pollinator. More detrimental to orchids are the antipollinators—spiders and insectivorous insects that remain on or among the flowers and seize potential pollinators when they approach the flowers (Ospina, 1969). One of the most noteworthy antipollinators is the Malaysian mantid *Hymenopus coronatus*. The young of this species may be either white, pink, or lilac, and the insects change color to match their substrate. They are found on many different orchids, as well as other sorts of flowers (Yong, 1976). We should also mention flower "robbers," which bite into nectaries and take the nectar without pollinating the flowers. Such robbers may make the visits of legitimate pollinators unprofitable. Bumblebees and carpenter bees are known to be robbers, though usually not of orchids.

SELF-COMPATIBILITY AND SELF-POLLINATION

Many kinds of plants are self-incompatible, though each plant is quite compatible with other individuals of the same species. This system requires cross-pollination and prevents inbreeding. Most orchids are self-compatible, but the flower structure favors outcrossing. This would seem to be the most flexible system, as an isolated plant still has at least the possibility of self-pollination and reproduction. In self-incompatible species, at least two individuals must invade a new area simultaneously in order to become established. Stoutamire (1975) reports self-incompatibility in *Cryptostylis*, and self-incompatibility is reported for a number of different species of *Oncidium*. Dodson (pers. comm.) has noted cases in which a single insect has pollinated a number of flowers on a single *Oncidium* inflorescence. The first pollination, which was accomplished with a pollinarium from another plant, results in a capsule, and all the other flowers fall off within several days.

After dwelling on the infinite complexities that promote cross-pollination in most orchids, one is almost shocked to learn that some populations are regularly self-pollinated (autogamous). These populations trade off the advantages of cross-pollination for the short-term advantage of producing huge numbers of seed. In some cases these flowers self-pollinate without the bud ever opening (cleistogamy). In orchid self-pollination, the rostellum may function as part of the stigma, the pollinia may fall or twist down onto the stigma, or supernumerary anthers may occur which are not separated from the functional stigma by a rostellum. Self-pollination may be temporarily favored by natural selection, especially if it is combined with a very adaptive genotype. Such plants dependably produce millions of seeds and may be real weeds. *Caularthron bilamellatum* and *Spathoglottis plicata* are self-pollinating plants that may be locally very abundant.

Self-pollination has a clear advantage when a plant species is extending itself beyond the geographic range of its usual pollinator. Thus, we find a number of self-pollinating forms in the West Indies and Florida that are represented by normal, outcrossing forms on mainland tropical America. It is noteworthy that very few species are 100 percent autogamous. There usually remains the possibility of some outcrossing. In *Cattleya aurantiaca* and *C. patinii* it is well known that some forms are always autogamous, or even cleistogamous, others are never autogamous, and others may be autogamous or not, depending on environmental variables. In one area in Chiapas, Mexico, where I found both *Epidendrum nocturnum* and *E. latifolium*, I saw only cleistogamous flowers of *E. latifolium*, but I found two plants that had all the earmarks of being hybrids between *E. latifolium* and *E. nocturnum*. Obviously, *E. latifolium* occasionally produces open flowers capable of cross-pollination. Self-pollinating flowers can function quite well without many of the usual orchid features, and mutations with petal-like lips, distorted columns, and so on may persist and multiply when they are associated with self-pollination. It is common to find flowers with three fertile anthers in self-pollinating populations (*Encyclia cochleata*, *E. boothiana*, *Psilochilus physurifolius*, for example). Self-pollination is frequent in saprophytes and is much more frequent among the more primitive orchids than in the advanced groups.

In apomixis, plants forego sexual reproduction entirely, and the embryos are genetically exact copies of the mother plant. This has been reported in *Spiranthes cernua, Dactylorhiza maculata, Zeuxine sulcata, Zygopetalum intermedium* (as *Z. mackayi*), and a few other orchids. This, like autogamy, is rarely obligate, at least among the orchids, and in some cases may occur only when the flowers are not pollinated or have been pollinated with pollinia from another species.

PSEUDOPOLLEN AND WAX

Orchid pollen is not used for food by flower-visiting insects, but a number of orchids offer pseudopollen, or a mealy, pollenlike foodstuff. We find such mealy food on the lips of most *Polystachya* species and some species of *Maxillaria* and *Eria*. This pseudopollen is gathered and used as food by insects, usually bees.

A few species of *Maxillaria* produce wax on the callus of the lip. Bees have been observed gathering the wax (Porsch, 1909), which may be used in nest construction. We have no detailed information on the pollination of these orchids.

EMPTY PROMISES

Several orchid flowers appear to offer food, but none is to be found there. Darwin found the spur of *Orchis* to have no nectar, and he believed that the insects were piercing the cells of the spur to get cell sap. More recent workers have found no indication that this is happening (van der Pijl and Dodson, 1966). Rather, it seems, the occasional inexperienced bee—and there are lots of inexperienced bumblebees in springtime—seeks nectar in the spur and then gives up, but usually repeats the experience at a few other inflorescences. Van der Pijl and Dodson report false nectaries on the lip of *Odontoglossum kegeljani*. In this case, the bee has great difficulty in reaching the seeming nectaries, and so it is likely to try several times before it is conditioned to avoid the flowers. Other orchids have a crest of yellow hairs that look very much like a cluster of pollen-bearing anthers. Thus, *Calypso* and *Arethusa* apparently depend on deceiving inexperienced bees. The same is true of *Calopogon*, but in this case the lip is hinged, and the bee is flipped against the column as soon as it lands on the imitation pollen (fig. 4.12).

MOVABLE PARTS

Movement in a plant is always surprising, even though it may be a mere twitch as compared with animal activity. Many orchids such as *Bulbophyllum* and *Peristeria* have hinged lips that move easily and are tipped

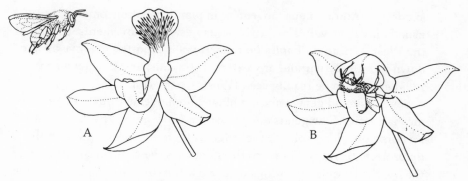

Figure 4.12 Pollination in *Calopogon*. *(A)* Flower in natural position, with the yellow callus simulating a cluster of pollen-bearing anthers. *(B)* The weight of the bee causes the hinged lip to fall, throwing the bee against the column.

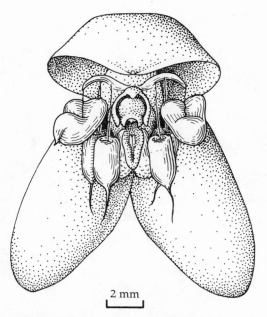

2 mm

Figure 4.13 Flower of *Bulbophyllum macrorhopalon*(?), showing the three mobile appendages attached to each petal. These move freely in the slightest breeze, so that the flower appears always to be in motion.

toward the column by the weight of the pollinator (fig. 4.10). These are merely passive movements, of course. Similarly, some species of *Bulbophyllum* and *Pleurothallis* have hairs or appendages that are moved by the wind (fig. 4.13). These seem to be especially effective in attracting the attention of flies and are a common feature of the fly-pollination syndrome. Autonomous movement in plants is not uncommon, but usually involves very slow, twice-a-day "sleep" movements. The lip of the West African *Bulbophyllum recurvum* is reported to move up and down repeatedly without any external stimulus, and the movement is rapid enough to be readily seen (Westra, pers. comm.).

Several orchid genera show a hinged lip movement that is by no means passive. The genus *Porroglossum*, for example, has a "sensitive" lip borne on a long column foot (fig. 4.14); in its normal position the lip projects forward, but if anything touches the callus near the base of the lip, the lip quickly springs up against the column. The insect is then enclosed between the column and lip, and the only exit is by the stigma and rostellum, where it is likely to remove or deposit pollinia. After

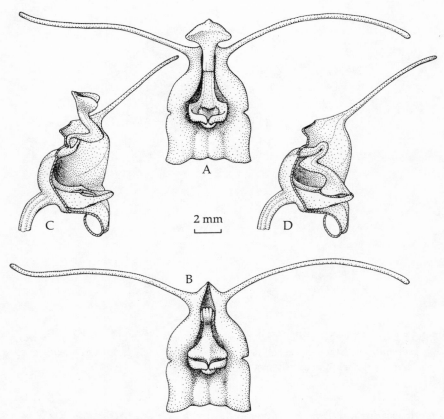

Figure 4.14 *Porroglossum amethystinum*, showing the sensitive, mobile lip. *(A)* Flower in the "ready" position. *(B)* Flower after the movement has been triggered, with the lip against the column. *(C)* Longitudinal section of a "ready" flower. *(D)* Longitudinal section of a triggered flower.

twenty minutes or more, the lip slowly returns to its original position and is ready to be sprung again (Sweet, 1972). Similar mechanisms occur in *Acostaea* and in the Australian genera *Drakaea* and *Pterostylis* (Northen, 1972). *Plocoglottis*, too, has a sensitive lip, but in some species it is a one-time reaction, and the lip cannot be reset once it is triggered (Holttum, 1955b).

In the Catasetinae one is impressed by the sling-shot ejection of pollinaria, though this is really a passive movement that results when tension is released. In the staminate flower of *Catasetum*, the stipe is stretched over a knob formed by the column (Northen, 1952; Dodson, 1962a). The anther and pollinia are located above the knob, and the viscidium is tucked in below the knob. As the flower matures, tension builds up in the elastic stipe. When the flower is fully mature, the slightest touch on the antennae that are borne on each side of the viscidium causes the viscidium to be released, and the whole pollinarium is thrown out with considerable force. The viscidium has a remarkably quick-setting glue, and the pollinarium remains firmly attached to whatever it hits (usually the back of a bee or the finger of a hobbyist). The other genera of the Catasetinae all show similar mechanisms, though the release mechanism is different in each genus and the degree of movement is less in some than in *Catasetum*.

TRAP FLOWERS

This rather self-explanatory term is another of the functional categories, like flag flowers and gullet flowers. The spring traps of *Porroglossum* and *Pterostylis* and the passive flip traps of *Bulbophyllum*, *Anguloa*, and *Peristeria* are all, of course, traps of a sort. But one of the primitive subfamilies of the Orchidaceae, the Cypripedioideae, has quite a different sort of trap flower. In this group the lip forms a pouch, or slipper, and access to this chamber is easy in front of the column. However, it is difficult to retreat through the same opening. Once an insect has entered the trap, it finds it much easier to exit to one side of the column. These exits are much smaller, and in leaving, the insect normally brushes against first the stigma and then the anther. Thus, it will pick up pollen, and on its second and successive visits to such flowers it will leave some of the pollen on the stigma. The species of *Cypripedium* are visited by various bees and seem to lack any obvious reward. Perhaps these depend on the inexperienced bees that are foolish enough to fall into the same kind of trap at least twice. The tropical lady slippers show many features of the fly pollination syndrome.

Another sort of trap flower is the flower of *Coryanthes* (see fig. 4.10).

Though the bees are not really enclosed, smaller bees are prevented from flying by being wet, and larger bees have too little space between the slippery walls to spread their wings and fly.

MIMICRY

Orchid hobbyists sometimes speak of mimicry in reference to all sorts of chance resemblances. In biology, however, mimicry refers to cases in which natural selection has favored a resemblance between individuals of different species. This is best known in insects, where palatable mimics may be almost indistinguishable from distasteful "models" of different genera or even different orders. We have already mentioned the mimicry between *Bulbophyllum* flowers and rotten meat, and the mimicry between *Calopogon* or *Calypso* and masses of pollen. There is growing evidence that some *Oncidium* species mimic the flowers of Malpighiaceous vines (Nierenberg, 1972). Female bees of the genus *Centris* gather oil from the sepaline glands of the Malpighiaceae. These same bees have been seen to seize *Oncidium* flowers for an instant and then fly away. Presumably, the bees are deceived and leave the *Oncidium* when they find no oil, but pollinate other plants when they make the same mistake again. The genus *Dracula* shows a very interesting mimicry. The flowers are borne near the substrate, the lip is remarkably fungus-like in structure, and the flowers often have a fungus-like odor. These flowers are visited by small flies, and Vogel (1978) suggests that the pollinators are fungus gnats, which seek to feed or lay their eggs on fungi. Vogel suggests a similar mimicry for *Corybas* and *Cypripedium debile*.

Van der Pijl and Dodson describe the Ecuadorian orchids *Epidendrum ardens*, *Elleanthus aureus*, and *Odontoglossum retusum*, whose flowers closely resemble the flowers of an Ericaceous shrub, *Gaultheria*, that is abundant in the same region. Similarly, Schelpe (1966) describes the resemblance of *Orthopenthea falcata* to the flowers of the sympatric shrub *Adenandra* (Rutaceae). Such resemblances are very suggestive of mimicry, as is the resemblance between the flowers of *Epidendrum radicans*, *Lantana camara*, and *Asclepias curassavica* (which are sympatric in Central America). Convergence between a scarce flower and a common one may be favored by natural selection, even if both species offer adequate rewards. Potential pollinators might ignore an infrequent flower which has a distinctive "search image" but fix on the search image of a common species.

A special type of near-mimicry has been characterized by van der Pijl and Dodson as pseudoantagonism, though the antagonism seems to be real enough. The females of the bee genus *Centris* often feed and gather pollen or oil from various flower clusters. The males tend to pick such flower clusters as the focal point for their "territories." These males then defend the territory against any other flying insect that is not a female *Centris*. When the wind moves the inflorescence, *Oncidium* flowers apparently look enough like an insect in flight to arouse the aggression of the male *Centris*. It then strikes the flower and receives pollinia on its face. Presumably the territory shifts from day to day, so that the bee is likely to repeat the behavior at another inflorescence. Perhaps this relationship should be described as "pseudotrespassing."

PSEUDOCOPULATION

This refers to cases in which orchid flowers mimic female insects and are pollinated by male insects in search of mates. What is truly remarkable about this pattern is that it has evolved several times, in different parts of the world with quite different groups of orchids and insects. The best known and most studied case is that of the European and Mediterranean genus *Ophrys*, whose flowers mimic female insects in form, color, odor, and surface texture (Kullenberg, 1961; Kullenberg and Bergström, 1976). The flowers mimic several different bees and wasps, and in spite of their remarkable mimicry, are not wholly species-specific. Thus, interspecific hybrids in *Ophrys* are fairly frequent.

It is interesting to contrast this with two other cases that are relatively well known. In Australia several different species of *Cryptostylis* are all pollinated by the males of *Lissopimpla semipunctata*, an ichneumon wasp. These *Cryptostylis* species are quite intersterile, and no hybrids are known (Stoutamire, 1975). In the same continent, there are a number of species of small wasps in the subfamily Thynninae (family Tiphiidae). The females of these wasps are wingless; they crawl to the tops of herbaceous stems and produce a sexual perfume, or pheromone. The winged males pick up the females and carry them to flowers, where they feed and mate. Species of *Caladenia*, *Chiloglottis*, and *Drakaea* all mimic these wasps, producing the appropriate odor and usually bearing dark, shiny calluses, which resemble the female wasp, on the lip (Stoutamire, 1975). In this case, each species of orchid mimics a different species of wasp and, again, there are few or no natural hybrids. The genus *Calochilus* is pollinated by wasps of the genus *Campsomeris*, and recent observations suggest that this, too, is a case of pseudocopulation (Jones and Gray, 1974). In the American tropics, orchids of the genus *Trichoceros* mimic tachinid flies and are pollinated by sexually aroused males. The related genera *Telipogen* and *Stellilabium* may also prove to be pollinated in this way. Other orchids have been suggested as being pollinated through pseudocopulation, but none of the other cases has been well documented.

Some of these cases of mimicry are quite remarkable, but we must remember that insects are not really very "intelligent," in spite of their complex behavior. Thus, a complex mimicry could start by the chance occurrence of a shiny surface and a pheromone-like odor. Once the pattern is started, natural selection can perfect this resemblance to the point that it is obvious even to our eyes.

Conservation of Orchids

Within the last decade, people have become very conscious of ecology and conservation. Rational evaluation is sometimes hampered by the emotion that colors practically anything said about conservation. Yet conservation is much too important for us to be irrational about it. We must consider all aspects of the problem and try to make our efforts as effective as possible.

COLLECTING

With respect to collecting orchids, we may offer a few commandments, mostly self-evident: One should never collect orchids in parks or nature reserves. One should not collect many plants of any one kind (except in rescue operations). One should never collect plants which cannot be grown in his garden or climatic area; many terrestrial orchids are so difficult to grow that they should not be collected except in rescue operations. One should never tell a commercial collector about a good orchid locality.

Most of us have a touch of the pack-rat instinct, but there is really no point in collecting more plants than we can provide a home for. When, however, we find an area that is being logged or otherwise destroyed, we should do everything possible to collect the orchids and distribute them to different growers, or even transfer them to a suitable habitat in a protected area.

COMMERCIAL EXPLOITATION

The commercial collector is often classed as an absolute villain, but in many developing countries the governments will give no money for the protection of orchids or butterflies unless they can see some profit in it. In all tropical countries large areas of forest are being destroyed annually by logging or agricultural activities, and most of the orchids in these areas are burnt. It would be far better to get some financial profit from them, and allow some of them, at least, to be maintained for a time in cultivation.

Another commercial venture that has given nature greater value in the eyes of some government administrators is the nature tour, a phenomenon of recent years that often represents a significant benefit to a developing country. Tours of orchid lovers, bird watchers, and other nature lovers from the United States and Europe to many parts of the tropical world are profitable for all concerned, when proper guidance is provided.

LEGAL PROTECTION OF ORCHIDS

Many countries have passed laws for the protection of some elements of their flora. Such legislation seems desirable in theory, but in practice the laws are often ineffective, and sometimes do more harm than good, as when they prevent rescue operations in habitats that are being destroyed. In too many cases, a well-meant law has little effect on vandals or commercial collectors, but makes life more difficult for honest plant lovers and botanists who try to comply with the law. It is nearly impossible to effectively protect small plants. Many epiphytic orchids may be carried out of the forest in one's pocket without making any appreciable bulge. Terrestrial orchids are easily pulled up, and may be pulled up merely because they are orchids. The evidence is easily discarded.

In the orchid world the most talked of law is the Convention on International Trade in Endangered Species of Wild Fauna and Flora (CITES). This treaty, which came into being in 1973, seeks to regulate international trade in endangered species or the products thereof. It includes three appendices: Appendix I, species threatened with extinction; Appendix II, species which could become threatened if trade were not regulated; and Appendix III, species not threatened by extinction, but listed by some signatory country in order to regulate trade in the indicated species. According to the convention, importation of species in Appendix I requires special permits from both the country of origin and the country of destination, while importation of species in Appendix II requires only a special permit from the country of origin (in addition to any permits that the importing country may require for any plant or animal imports).

The pressures which led to the development of the convention grew out of concern for wild animals and their products. Plants were grafted into the treaty late in its development and seem to be an afterthought. Rather than doing the necessary investigation to determine which orchid species to include, the authors of the convention simply put the whole family in Appendix II. This is patent nonsense, of course, for over 90 percent of the orchid family have very little commercial value and will never be threatened by trade. The argument was, apparently, that customs agents could not be expected to distinguish different orchid genera or species, yet this same treaty lists mammalian subspecies which even expert zoologists could scarcely distinguish without knowing the exact geographic origin. There are, no doubt, look-alikes, such as the one-leaved *Cattleya* and *Laelia* species, and it would be reasonable to include all of them in the protected list in order to protect those that are really threatened by extinction. But putting fifteen to twenty thousand species which do not need this type of protection into Appendix II has the same effect as putting unnecessary stop signs all over town: it weakens the credibility and enforceability of the law. Changing the treaty, however, will be a hard task, and we may have to live with its absurdities for a very long time. In theory, the United States will have "rescue centers" to take over the care of endangered plants that are confiscated because of a lack of proper permits. In practice, the funds are not now available for such centers, and they would be difficult to fund, even if the appendices were limited to the species that really need protection. As a result, thousands of orchid plants have been burnt in the name of conservation.

Some orchid groups are truly threatened by commercial collectors, the species of the genus *Paphiopedilum* being an outstanding example of this, though none is among the nine orchid species currently listed in Appendix I. Still, there are less than fifty orchid genera that are so horticulturally desirable as to be in any way threatened by trade. Even if the convention were effectively enforced at the international level, it still gives little protection within the countries of origin, where commercial collectors continue to bring plants into the cities for sale, dooming most of them to a lingering death.

Orchid societies could contribute to the conservation of orchids by organizing a survey that will determine which orchid species are truly endangered or likely to become endangered. Such a study would be very useful for all phases of conservation.

CONSERVATION THROUGH CULTIVATION

With the rapid destruction of natural habitats and the great interest in cultivating orchids, it is only natural to try and save orchid species in gardens. The potential here is great but is not without its problems. An individual can do a great deal—for a few decades—but conservation is a long-term problem, and continuity is critical. For this reason, botanical gardens and orchid societies have an advantage, but even here success depends on individual action. To maintain an effective conservation program, a botanical garden must have a sympathetic director, an able and willing grower, and, of course, funds. For tropical orchids, botanical gardens in the tropics are more practical, in that orchids can be grown with less work and less investment, but some long-term interest and investment is necessary. A few botanical gardens at different elevations in the tropics could accomplish a great deal, but there are not now many that are assured of long-term support.

To effectively cultivate rare species, they need to be propagated and distributed to different gardens and different growers as much as possible. In cases of extreme rarity, we may self-pollinate, or even meristem a plant, but this reduces the genetic diversity of the "population," and a clone contains very little genetic diversity, indeed. In fact, in a self-incompatible species, a thousand plants of the same clone have little real advantage over a single plant; they cannot be self-pollinated or crossed with each other. It is far better to cross-pollinate, using different individuals of the same species. Alphonso (1976) points out how crossing and selection of *Vanda coerulea* and *V. sanderiana* have produced beautiful populations very unlike the natural species—plants which might not survive in the wild. Alphonso favors self-pollinating, but that is really a poor policy.

Reintroduction of scarce species into their natural habitats is a desirable goal, but this must be done very carefully and only in the correct area. Species that might hybridize with other related species in an area should not be introduced there, since this could cause both populations to lose their identity, and could cause extinction in a rare species. We really know very little about the best ways to reintroduce species into nature.

One difficult question is which are the rare species that we should maintain in cultivation? There are not enough botanical gardens in the world to accommodate all of the plant species that are or may become threatened with extinction. Further, the fact that a species is rare in cultivation does not mean that it is not common somewhere in nature. Probably, as suggested by Arp (1977), the major role for the botanical gardens (and private efforts, as well) is to propagate the rare species of some horticultural interest, thereby taking some of the pressure off of the natural populations. It is to be hoped that botanical gardens and orchid societies will plan and coordinate their conservation activities so that the available resources will be used most effectively.

CONSERVATION OF HABITATS

All of the approaches we have mentioned can help in the conservation of orchid species, but the single most important thing is the conservation of habitats. The epiphytes, especially, will survive very well in nature as long as their habitats are protected. One sees a striking example of this in Venezuela. *Cattleya mossiae* thrives within a stone's throw of Caracas, because the forest in Parque del Avila is effectively protected. At the same time, *C. lawrenceana*, *C. jenmannii*, and *C. violacea* are all endangered by habitat destruction and commercial collection within the country, even though they are native to some of the most remote areas in Venezuela.

We cannot hope to conserve all good orchid habitats; the pressures for economic development are much too great. But there are many good reasons for preserving forested areas, including the continued survival of *Homo sapiens*. We must try to preserve representative habitats wherever we can. For the conservation of orchids, rugged, mountainous terrain has several advantages. Its varied topography offers appropriate sites for a greater variety of orchid species, and many of them will be sufficiently inaccessible to discourage illicit collectors or vandals. Then, too, such areas are much easier to set aside than good agricultural land. Our reserves should be as large as possible. Very small reserves will eventually lose many species that are normally thinly scattered. Then, too, a small reserve may be inadequate to maintain some of the host tree species or some of the pollinating insects without which the orchid species cannot survive. If the reserves are large enough to have some natural disturbance from time to time, management may not be necessary. In practice, though, most reserves need to be managed, especially to maintain populations of some terrestrial orchids that are a bit weedy in their requirements. Controlled burning, mowing, grazing, or tree-felling may be needed to keep appropriate habitats available. Finally, I must emphasize that the legal establishment of a park or reserve is only the first step. The area must be effectively protected, and this requires funds for rangers and patrols.

Evolution

5

The Orchidaceae seem to be a family in a state of active evolution. We find "good" species, semispecies, and variable complexes, just as one would expect of a group in active speciation. "Problem genera" abound, and we find almost diagrammatic phyletic trends within and between groups. While we should not try to derive any living group from any other, we cannot but notice that the family has many seeming links that tie groups together. Disagreements about the classification of subfamilies and tribes within the family are not entirely due to bad taxonomy. One of the interesting things about the family is that we can usually be sure in which "direction" our phyletic trends are going. At one end we have slightly odd lilies, at the other end, structures that practically transcend the concept of flower, and vegetative habits that are unique.

Naturally, different groups are "advanced" in different features, and often both specialized and relatively unspecialized features are found in a group at the same time. This is as typical of the family as it is of the flowering plants as a whole. A consideration of phylogeny and relationships within the family supports very nicely the concept of "genetic uniformitarianism" (Stebbins, 1974). What seems a good tribal or subtribal feature in one group will differ between closely related species in another group. This should not be surprising if we believe in evolution. Only in the case of planned, special creation should we expect a neat hierarchy of family, subfamily, tribal, generic, and specific features. In such a case, taxonomy would be no challenge, and phylogeny nonexistent.

Reproductive Isolation

In order for two closely related species to occupy the same habitat and yet maintain their identity as distinct species, there must be some sort of barrier or reproductive isolating mechanism. Indeed, reproductive isolation is considered a critical step in speciation. Some botanists have become disenchanted with reproductive isolating mechanisms because they do not always correlate with what we see as species. Still, we may be sure that any two species that coexist without losing their identities have some sort of isolating mechanism. No feature can, by itself, be an isolating mechanism; it is only in the interaction between populations that isolating mechanisms exist, and there are usually several different features that together form a genetic barrier between any two populations (Levin, 1978).

Of the many different classifications of isolating mechanisms, Levin's system (1978), with modifications, will be used for our discussion of orchids. We find, contrary to some reports, that most or all of the types of barriers that Levin lists are represented in the Orchidaceae, and one must emphasize all of these barriers have a genetic component.

1. Prepollination mechanisms
   A. Temporal
      Seasonal
      Diurnal
   B. Floral
      Ethological
      Structural
      Chromatic
      Olfactory
   C. Reproductive Mode
2. Postpollination mechanisms
   D. Incompatibility
   E. Hybrid unfitness
      Hybrid not viable in nature
      Hybrid sterile
      Hybrid breakdown in later generations

Pollination seems to be an especially critical stage for reproductive isolation among orchids. In a sense, prepollination mechanisms are "better" than postpollination mechanisms, in that the plant is not obliged to invest energy in the development of hybrid seed, nor do foreign pollinia prevent the development of nonhybrid seed.

There may be advantages to what Stebbins (pers. comm.) has termed a "leaky barrier," that is, a barrier that is strong enough to maintain the identity of the species but still lets in a bit of gene exchange now and then to add to the gene pool.

TEMPORAL BARRIERS
Many sympatric species such as *Cattleya skinneri* and the very similar *C. patinii* are effectively separated by their different flowering seasons.

I know of no orchids that are isolated from each other by flowering at different times of day, but we do find ephemeral orchids which flower on different days in response to the same environmental cue, a barrier that lies somewhat closer to diurnal than to seasonal. Further, we find many species whose flowers are fragrant only for a short time each day (or night). In general, though, orchid pollinaria can remain on the pollinator for hours, if not days, so that a difference of a few hours is unlikely to form an effective barrier.

FLORAL BARRIERS

Here the orchids offer a wealth of material. I follow Levin in listing ethological (behavioral) mechanisms, but in fact these can scarcely be considered separately, since differences in pollinator behavior always arise in response to features of the flowers. Except for extreme cases of mechanical isolation, there is an element of both behavior and structure (or chemistry) in all interactions between pollinator and plant. One such extreme case involves the large and small gullet flowers. The pollinator of the larger flower could not enter the smaller one, and the smaller pollinator could not effectively pollinate the larger flower, even if it could reach the reward. We find a series of small-, medium-, and large-flowered species in the larger genera that have gullet flowers, such as *Cattleya, Cymbidium, Eulophia,* and *Sobralia.*

Differences in spur length may be important, as in *Platanthera psycodes* and *P. grandiflora* (Stoutamire, 1974a). Color is not usually a decisive isolating mechanism in itself, though it may be important in Lepidopteran pollination, as in the orange-flowered *Platanthera ciliaris,* pollinated by butterflies, and the white-flowered *P. blephariglottis,* pollinated by moths (Smith and Snow, 1976). Color also seems to be critical in pollination by birds. Perfume, like color, is usually a contributing factor, but in euglossine pollination and in pseudocopulation odor is decisive in determining which pollinator visits a flower.

REPRODUCTIVE MODE

Any plant population whose members are self-pollinating is, at once, isolated from its relatives. Cleistogamy, if absolute, would establish a firm barrier against hybridization with other populations.

INCOMPATIBILITY BARRIERS

While incompatibility between closely allied species is not the rule in the Orchidaceae, it is by no means unknown. The species of *Cryptostylis,* for example, are incompatible (Stoutamire, 1975), and in the large (and somewhat unnatural) genus *Oncidium* there is a high degree of incompatibility between several groups (Sanford, 1964). Information on this subject is generally unsatisfactory. We have extensive records of the crosses that are successful but usually none of the failures, nor any record of the number of crosses needed to get a "take" (see, however, Moir, 1975b).

HYBRID UNFITNESS

Similarly, we have little data on the viability of hybrids in nature, though it is probable that some interspecific hybrids fail to find a niche in nature and that other viable ones fail to attract an effective pollinator. Hybrid sterility does occur, though here again the data are unsatisfactory. *Epidendrum ibaguense* and *E. calanthum* are so similar that some taxonomists consider them synonymous, yet I have found the artificial interspecific hybrid to be consistently sterile. Many of the interspecific and intergeneric crosses that have been made artificially are sterile, especially when there are differences in chromosome number between the parent species. The intensive efforts to produce hybrids in certain combinations obscure the hybrid breakdown that must be overcome (see, for example, Burns, 1961; Tanaka and Kamemoto, 1961; and the several articles by Moir).

Some authors imply that interfertility is the rule within subfamilies, but in fact interfertility usually occurs only within subtribes. In the great majority of these cases the species are completely isolated in nature by prepollination barriers. Even if these barriers were breached, the hybrids would scarcely be viable under natural conditions and would have definitely reduced fertility. When incompatibility occurs between closely related species, it probably has been favored and perfected by natural selection, as in the case of *Cryptostylis*, where several species are using the same pollinator. While this sort of incompatibility is infrequent in the orchids, the incompatibility that arises as a by-product of genetic divergence in more distantly related taxa is the rule.

Breeding Systems

Modern biologists recognize that a plant species' genetic system is dynamic and adaptive, as subject to natural selection as shape of leaf or color of petal. Baker (1959) and especially Grant (1958) delineate an extensive spectrum between obligate outcrossing and obligate self-pollination (see also Fryxell, 1957). Though orchids are only infrequently obligate outcrossers, they are nearer the outcrossing end of the spectrum. Their floral mechanisms, the longevity of the plants, and their relatively high chromosome numbers are features favoring recombination and continuous variation within the population. Further, the genetic isolating barriers are often leaky enough to augment the gene pool, as is usual in perennial plants with complex floral mechanisms (Grant, 1971). It is interesting that *Psygmorchis,* which has a relatively short reproductive cycle, also has the smallest chromosome numbers yet reported in the family (2 $n = 10, 14$), suggesting that selection has favored a reduction in recombination in these short-lived micro-orchids. Unfortunately, we know very little about chromosome number in micro-orchids, but the report of 2 $n = 24$ for *Taeniophyllum aphyllum* is also suggestive.

In the earlier stages of flower evolution, when pollen was the main reward, there was surely intense selection pressure for self-incompatibility, for in that system the stigma has to be borne among the anthers, and there is no structural way to avoid self-pollination. With the development of more complex floral mechanisms, this factor has become less important, and some groups can maintain a high degree of outcrossing without self-incompatibility. Such systems are more flexible and much more suitable for long-distance dispersal, since a self-incompatible species can become established in a new area only if at least two different individuals invade the area at the same time.

Pollination Mechanisms as an Evolutionary Theme

Biologists now recognize a general evolutionary trend from nonspecific, promiscuous pollination systems to more restricted, highly specialized systems. The most primitive orchids (both real and hypothetical) are already well toward the specialized end of this spectrum. Where have they gone from there? Unlike some other groups, they have not, to any very great extent, adapted to pollination by social bees. In northern regions, and sometimes at higher elevations, there is considerable dependence on bumblebees, and both honeybees and stingless bees are known to pollinate a few orchids, but the social bees make up a small part of the pollination spectrum. One reason may be that the social bees are poor pollinators for widely dispersed populations. In many cases the orchids have become specialized to particular groups of nonsocial Hymenoptera; they have radiated especially to Lepidoptera and Diptera, groups which these precision flowers can use more efficiently than most other flowers, and they have often evolved away from a food reward to what we may call nontrophic pollination systems.

The adaptive radiation to different pollinators and different systems of pollination occurs on several levels. Vogel (1954) documents the intrageneric adaptive radiation of *Disa, Satyrium,* and *Mystacidium* to different types of pollinators. Different pollination systems often correspond rather well to generic boundaries. In the Laeliinae, for example, *Cattleya* and *Laelia* appear to be large-bee flowers, and there is not a very effective isolating barrier between these supposed genera (referring only to *Laelia* section *Cattleyodes*). *Brassavola*, on the other hand, is a moth flower, and intergeneric hybrids between this genus and *Cattleya* or *Laelia* are very rare. *Encyclia* appears to be basically adapted to small bees and wasps. *Epidendrum* is primarily adapted to Lepidoptera, though there has been some radiation to flies and hummingbirds. The small genera *Neocogniauxia* and *Hexisea* are surely hummingbird flowers, but some of the members of *Hexisea* may have evolved from the closely related *Scaphyglottis* independently of each other, in which case *Hexisea* would be an artificial genus.

NONTROPHIC POLLINATION SYSTEMS

In reviewing the ways in which orchids are pollinated, one is struck by the number of systems in which the orchid offers no reward (relying on mimicry or deceit) or offers a reward other than food (perfume or wax). It seems unlikely that the energy saved by producing no nectar is important enough to explain this pattern. What, then, are the ecological and evolutionary reasons for this stinginess on the part of the orchids? There are several ways in which some or all of the nontrophic pollination systems offer advantages to the plant over the basic system of food as reward.

Better isolating mechanisms. A slight morphological or olfactory difference may constitute a firm isolating mechanism in a pseudocopulation system, while it is unlikely to do so in a food-based system (Proctor and Yeo, 1972). The same is true in pollination by euglossine males, where the addition of a single component to the fragrance may constitute an effective barrier. Both of these pollination systems are highly specific and seem to offer maximum opportunities for speciation.

Greater fidelity. Pseudocopulation and the production of euglossine-attracting terpinoids impose a high degree of fidelity, almost a compulsion. In each case it is very probable that the insect will revisit the same species if the opportunity is presented. The offering of food as a reward can impose fidelity only if the flower is reasonably common and the visits are energetically worth the effort (Heinrich and Raven, 1972).

Pollination over greater distances. Again, pseudocopulation and production of terpinoids seem well suited for pollination in a dispersed population. The travels of food-seeking females are circumscribed to the areas and the routes that are productive of food; males, on the other hand, have no home base and are more inclined to wander randomly in their search for a scarce resource, virgin or willing females. In the case of euglossine pollination, the evidence is compelling that these perfumes attract the males over great distances, even though measurements are hard to obtain (Williams and Dodson, 1972).

Forest trees and large vines can use the "cornucopia" strategy of producing a large mass of flowers over two or three weeks in order to obtain a certain percentage of cross-pollination (Gentry, 1974; Frankie, Opler, and Bawa, 1976), but this is not a very efficient system. In any case, small epiphytes cannot produce enough advertisement and reward to divert many food-seeking bees from their regular routes. Much the same problems apply to the "steady state" strategy—producing one or few flowers daily over a long period. If two epiphytes of the same species were fortunate enough to be on the same "trapline," some crossing could be expected, but the epiphytes could scarcely offer enough of a reward to induce many detours.

Greater outcrossing. Several nontrophic systems seem to offer not more but less pollination than a food-based system, but they do achieve a higher percentage of cross-pollination. If an inflorescence offers food, the bee will work every flower in the inflorescence and probably self-pollinate most of them, unless there are structural devices to prevent this. If, however, the reward is imaginary, the bee is unlikely to stay long on any one plant. The same is true of pseudocopulation. Another way of preventing self-pollination, of course, is to bear only one flower, though this strategy runs the risk of reducing the advertisement.

Longer visits. This applies only to euglossine pollination, but the extended visits by the bees permit the evolution of some peculiar mechanisms that would be practically impossible under any other system. This also permits a good deal of adjustment in the system. Some of the simpler gullet flowers, like *Acineta*, seem at first glance to be very poorly designed. The flower "fits" the bee so loosely that pollinia are removed only occasionally. A simple tightening of the flower would surely increase the efficiency of removal and deposition, but it would also increase the percentage of self-pollination. I would guess that these systems are adjusted for something close to the optimum percentage of cross-pollination.

ADAPTIVE RADIATION WITH ONLY ONE BEE
Stebbins (1970) refers to "differential adaptations to the same pollinators," a process that, among orchids, goes far beyond one plant depositing pollen on the back of the pollinator and another depositing it beneath. This is especially clear in the euglossine-pollinated orchids,

which show a tendency for the commonest and most widespread orchid species to be pollinated by the commonest and most widespread bee species. This results in considerable selection pressure among the orchids to share individual pollinators, in spite of the number of euglossine species that occur in many areas. One can delineate thirteen different places where pollinaria may be attached to a bee (fig. 4.11), and of these, only the two that involve placing the pollinaria under the scutellum or between the thorax and the abdomen are likely to interfere with each other. I know of no one bee species on which all these sites are used (it would probably be grounded), but in the majority of cases when a single bee species is "shared" by sympatric orchid species, the pollinaria are placed differently on the bee (Dressler, 1968b).

Coevolution

Coevolution and coadaptation are fashionable words now, and as sometimes happens with such words, their meanings are not always clear. As I understand it, we may define coevolution as "a relationship in which two kinds of organism each influence the evolution or adaptation of the other." "Kinds" of organism is deliberately vague here, for it may refer to a local population within a species or to a whole pollinator class, such as butterflies. I emphasize the flexibility of the definition, for in too many cases authors seem to visualize one plant species and one insect species, petal in tarsi, marching down the aisle of coevolution and becoming ever more tightly adapted one to the other. I do not deny that such a scenario may happen. Figs and fig wasps are mutually interdependent and highly adapted to each other. The same is true of *Yucca* and *Tegeticula*, but both of these cases are exceptional. There are many oligolectic bees, but these are largely a feature of desert habitats and bear little relevance to orchid pollination.

In the vast majority of food-based pollination systems, the insect depends on many different plant species for nectar, pollen, resin for nest-building, and other resources. There is, of course, a certain degree of mutually directed evolution in bees and bee flowers, or in any other such combination, but it is one-sided. The bees are adapted to flowers as a resource, sometimes to a particular class of flower, but rarely (outside of deserts) to a single genus. Many euglossine bees are adapted to feed in deep, tubular flowers, but their nectar is found in Apocynaceae, Bignoniaceae, Convolvulaceae, Marantaceae, Rubiaceae, Zingiberaceae, and other families. In the nontrophic pollination systems, this situation is, if anything, even more one-sided. Euglossine males are well adapted to gathering perfume droplets from the surfaces of flowers, but none of them is adapted to any one type of flower. In all cases of mimicry, the adaptation is very much on the plant's side, and the mimic must keep its numbers within certain limits for the system to work. If the plant's interference with the insect's life becomes too great, there may well be adaptation on the part of the insect, but its effect will be to disrupt the relationship.

Speciation

We have already considered some aspects of speciation as they involve isolating mechanisms, but we can here consider speciation itself as the process which permits a population to go its own evolutionary way with little or no gene exchange with the parent population.

Gradual speciation in geographically isolated populations is usually considered one of the major patterns of speciation, or even the only one. This is, I am sure, important in the orchids, but few populations have achieved species status merely by accumulating genetic differences that cause incompatibility. With the breeding system of the orchids, this would require a long period of isolation indeed. Of course, in some cases differences in chromosome structure or number which have become fixed in an isolated population might impose reproductive isolation within a short time period. In most cases, though, we must demand that some prepollination isolating mechanisms develop. Simple differences in flowering time may sometimes be enough. More often, though, we would expect adaptation to a different pollinator to be the critical factor. In such a pattern of evolution, Stebbins (1970) stresses the necessity of a period when two different types of pollinator are both effective. We may point to the case of *Cattleya skinneri*, which has all the earmarks of a bee flower, and *C. aurantiaca*, which seems to be a hummingbird flower. While we do not know which agent is responsible, there is no doubt that some one organism still effectively pollinates both species, even if rarely, for they hybridize in southern Mexico and Guatemala to produce *C. × guatemalensis*. Quite a hybrid swarm occurs, and some elements of the complex have been named as species (*C. deckeri, C. pachecoi*).

Gradual speciation, however, is a well-known phenomenon, even if rarely observed from beginning to end. The isolating mechanisms in the orchids suggest that they also have fairly abrupt speciation. If an orchid invades a new area where its normal pollinator is lacking but a very different agent is minimally effective, we would expect intense selection pressure for a reorganization of the flower to accommodate the new pollinator. Much the same pattern could result from the disappearance

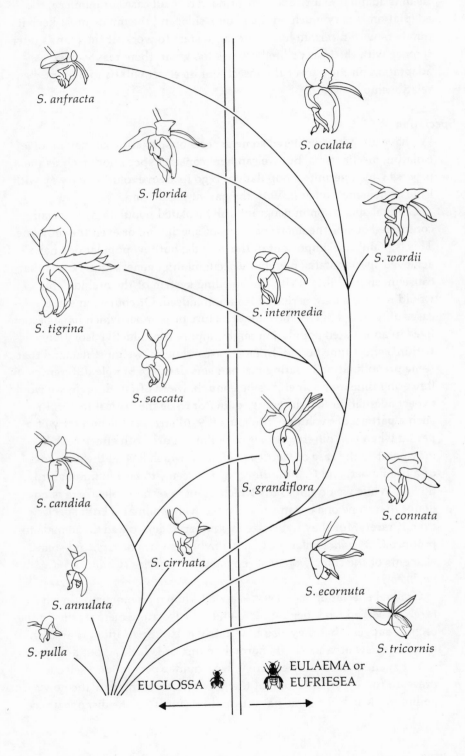

S. anfracta

S. oculata

S. florida

S. wardii

S. tigrina

S. intermedia

S. saccata

S. candida

S. grandiflora

S. connata

S. cirrhata

S. ecornuta

S. annulata

S. pulla

S. tricornis

EUGLOSSA

EULAEMA or
EUFRIESEA

of the normal pollinator in a portion of its range. In the euglossine bees we have about fifty species of large bees (*Eulaema, Eufriesea*) and about twice as many smaller ones (*Euglossa*). There is every indication that some orchid genera have "switched" from one to the other several times, in what Dodson (1962b) has called "leap-frog" speciation (fig. 5.1). The interesting thing about this process is that adapting to a much larger or much smaller bee may lead to a different physical mechanism of pollination, and possibly to the evolution of a new genus. A good example is *Paphinia clausula*, which appears to have "closed" its flower in order to achieve better pollination by rather small bees (Dressler, 1968a). In this case, the flower deposits pollinaria on the bees' legs rather than under the scutellum as in the other species of *Paphinia*. The improvement of this relationship could easily lead to a new generic pattern.

That pollination mechanisms may correspond to genera is beautifully illustrated by the *Chondrorhyncha* complex in which the genera were described and characterized without reference to pollination relationships (Garay, 1969). We have found that *Chondrorhyncha* places the pollinaria on the scutellum, while the closely related genus *Cochleanthes* places the pollinaria behind the head. *Kefersteinia* is unique in placing the pollinaria on the scape of the antenna, and the recently described *Chaubardiella* places its pollinaria on the trochanter of the bee's leg. Such correlations of pollination mechanism with character combinations that taxonomists have chosen for generic recognition is rather frequent in the euglossine-pollinated orchids (Dressler, 1976b).

Another fascinating possibility appears when we consider either pseudocopulation or euglossine pollination. Here, a single mutation which affects the odor of the flower could start the evolution of a new species at once and without geographic isolation. Similarly, in the case of pseudocopulation, a marked change in the form or color of the lip could attract a new pollinator or cause an established pollinator to behave in a different way. Stebbins and Ferlan (1956) suggest that interspecific hybrids in *Ophrys* may sometimes find a new pollinator and be stabilized as new and reproductively isolated species. The same thing is a possibility in euglossine-pollinated orchids. I have observed that the

Figure 5.1 Leap-frog speciation in the genus *Stanhopea*. The more primitive species of *Stanhopea* are adapted to pollination by small bees of the genus *Euglossa*. Many of the more advanced species are adapted to pollination by large bees of the genera *Eulaema* or *Eufriesea*, and some species, such as *S. anfracta* and *S. tigrina*, seem to represent a shift back to pollination by *Euglossa*. (Redrawn from Dodson, 1962b.)

hybrid between *Gongora gibba* and *G. quinquenervis* attracts bee species which do not visit either parent species.

Success or Failure?

One often reads speculation about whether or not a plant group should be considered successful, but (as far as we know) plants have no definable goals except to survive and reproduce, and so there is no other objective measure of success. Several different things may be implied when one speaks of a "successful" group of plants.

Population size. By this criterion, many orchid species are quite successful. There are some species that are highly dispersed, and others that truly seem to be scarce, but "rare" usually means "hard to find where we looked." Again and again, "rare" species prove to be frequent in the right habitats, and many epiphytes are exceedingly numerous, even though somewhat inaccessible by terrestrial standards.

Number of species. Here, clearly, the orchids are a successful group.

Morphological diversity. Again, the orchids are quite successful by this criterion.

Aspect dominance or biomass. Orchids rarely dominate the landscape, as do trees or grasses, and only in exceptional habitats do they make up a large proportion of the total biomass. If sheer bulk is to be our definition of success, then we must consider the orchids to be only moderately successful.

We frequently read that the orchids, or some group of orchids, are so overspecialized that they are doomed to extinction. For the most part, this is nonsense. The orchids are generally rather specific in their pollination relationships, but scarcely as much so as the thriving genus *Ficus*. If the general trend in floral evolution is toward greater specialization, can we reasonably slight the orchids for having gone a bit further than most other groups? Some orchids are, to be sure, strongly dependent on certain groups of insects, but they have usually "chosen" groups of insects that are thriving. If *Homo sapiens* succeeds in destroying the forests everywhere, then many orchids will surely disappear, along with other forms of life. But given a chance, they will continue to thrive. Who knows what new and unexpected lines of evolution they may take up in the next few hundred thousand years?

Evolution of Habit

The patterns of evolution in orchid growth habit have been well outlined by Rolfe (1909–1912). We believe that the primitive orchids were sympodial, with relatively slender rhizomes, fleshy roots but no storage roots, elongate stems, spirally arranged, nonarticulated plicate leaves, and a terminal inflorescence. We find this habit in most subfamilies, and it shows a strong correlation with other primitive features. This is the habit of *Neuwiedia, Cypripedium, Cephalanthera, Epistephium,* and *Palmorchis.* We can get a fair idea of the diversity that has evolved from the primitive growth habit by scanning the habit diagrams in figures 3.2 and 3.3. In the absence of a fossil record, though, we can only guess, on the basis of what we know of the family, about the major trends of evolution. The speculations I will present here fit well with all that I know about the orchids, though this outline must gloss over a multitude of minor trends, parallelisms, and detours. But if it prods someone else to learn more about orchids and their relationships, then it will have served a purpose.

The Apostasioideae are clearly a relic group, and both genera are close to the primitive habit.

In the Cypripedioideae, *Cypripedium* itself shows the primitive habit, though it also shows some modifications with fewer and wider leaves. *Selenipedium,* though very primitive in ovary structure, has distichous, rather narrow leaves. The main trend in the evolution of habit within this subfamily has been the aggregation of distichous, fleshy, conduplicate leaves. This trend probably occurred in steep, rocky habitats that formed the stepping stone to epiphytism; both *Phragmipedium* and *Paphiopedilium* are primarily humus epiphytes.

Among the Spiranthoideae we find the primitive habit well represented in the Tropidiinae. The Goodyerinae have taken up a habit that is unusual within the family and is primarily suited to wet forests: there is no sharp distinction between rhizome and aerial shoot, and the stems creep along the soil, in the leaf litter, with the leafy shoot more or less erect. There are a few epiphytes with fleshy roots. In the Cranichideae, on the other hand, a rosette habit with storage roots is the predominant habit, and there has been little modification. Here again there are a few humus epiphytes and even a few small plants that are nearly twig epiphytes (*Eurystyles, Lankesterella*).

Among the Orchidoideae, the northern Neottieae retain the primitive habit or, in *Listera,* a condensed version of it. The Diurideae, Orchideae, and Diseae, with well-developed tuberoids in most groups, include similar habits and have evolved many variations on the rosette plant theme, ranging from single, near circular leaves (*Bartholina*) to the onionlike leaves of *Prasophyllum.* Again, there are a few humus epiphytes and rock dwellers.

Up to this point, we have not encountered corms or pseudobulbs, but in the remaining subfamilies these are a recurrent theme. It seems that there have been two main patterns in the evolution of epiphytes, which we may call the reed pattern and the corm pattern (see fig. 3.2). In the reed pattern, there is strong selection for distichous, conduplicate leaves; the inflorescence may remain terminal, and pseudobulbs may develop as slender-based thickenings of several internodes. In the corm pattern, the base of the stem forms a corm or pseudobulb, the leaves are likely to remain plicate, and there is stronger selection for a lateral inflorescence. These two patterns represent only the beginnings of a complex radiation, and the further embellishments on both patterns are likely to show many parallelisms, all leading toward fleshy, conduplicate leaves, pseudobulbs of a single internode, and lateral inflorescences, or monopodial micro-orchids.

Among the epidendroid orchids we find the primitive habit in *Epistephium* and *Palmorchis*, especially. *Vanilla*, forming fleshy vines, clearly has gone off on a tangent not paralleled by most other orchids. Most of the Gastrodieae have gone off into saprophytism, with the drastic vegetative modifications that implies. *Nervilia*, however, has corms and thus resembles the Arethuseae and some Vandoideae. Among the Arethuseae, *Arundina* and the Sobraliinae represent the reed pattern of evolution, while most Bletiinae clearly fit the corm pattern, most having gone not much beyond the basic corm pattern. The Epidendreae fit the reed pattern of pseudobulb evolution, through the Bulbophyllinae have gone further in this line than the other groups, developing pseudobulbs of a single internode. Among the Epidendreae, the Pleurothallidinae have not developed pseudobulbs at all, having a single fleshy leaf and a slender stem. As we have learned to expect of this family, they have achieved great diversity with this simple pattern.

Both the Malaxideae and the Coelogyneae seem to belong to the corm pattern of evolution, though both retain a terminal inflorescence. In Asia the Malaxideae have gone far beyond this stage, producing many epiphytes with fleshy, conduplicate leaves. In the Coelogyneae, the pseudobulbs have been reduced to a single internode, and the inflorescence is often pseudolateral.

Among the Vandoideae, as with the Epidendroideae, both main patterns are represented. In the Polystachyeae, *Neobenthamia* has slender, reed-like stems and a terminal inflorescence, and this habit may well be primitive for the tribe. For the Vandeae, there is no reason to suppose that a corm or pseudobulb stage was ever present. The monopodial habit could easily evolve from the sympodial habit by the development of a lateral inflorescence and by retention of apical growth. In both the

Maxillarieae and the Cymbidieae, the cormous habit with plicate leaves is well represented among the more primitive members. In both tribes a bewildering array of habits has evolved from this basic pattern, with pseudobulbs of one or several internodes, or no pseudobulbs, plicate or conduplicate leaves, and, in the advanced groups of each tribe, mono-podial micro-orchids with fleshy leaves.

Evolution of Flower

The evolution of orchid flowers must have started with a generalized lily-like flower. Such a flower would have an inferior ovary, and the perianth would be more or less tubular (but not united). The flower was already resupinate, perhaps by merely bending down alongside the rachis, rather than bending away from it. The loss of the anthers on one side is analogous to the evolution of the typical gullet flower of the Scrophulariales and must have resulted from the same factors—those involved in pollination by bees or wasps (if indeed it was possible to tell one from the other at that stage). With the pollinator entering the flower always in the same position and under the style, it is clearly the dorsal anthers which will place pollen where it will be effective. Any change that suppresses the ventral anthers or moves them to a dorsal position should be of adaptive value. In the dicotyledonous Scrophulari-ales only a single anther was completely suppressed, and the other four moved together on the dorsal side of the gullet. In later stages, another pair might be lost, but the dicot gullet flower always has at least two anthers side by side, a difference which is critical in shaping further trends of evolution. In the case of the proto-orchids, three anthers are lost, or perhaps first one and then two others. At the same time the lower petal, which consistently functioned as a landing platform, doubt-less began to differentiate from the other two, becoming larger and perhaps developing callosities which would aid and guide the pollinator. With the loss of the three ventral anthers and the development of a lip, we could now consider these plants as primitive orchids.

The relationship between the stigma and the anther was quite critical. We have an analogous sort of monocot gullet flower in the Zingibera-ceae, but there the stigma projects beyond the anther. Such a flower may look vaguely like an orchid, but it could never become one. In early orchids the anthers projected somewhat beyond the stigma, so that in leaving the flower an insect could brush first the stigma and then the anther. Given also the saprophytic tendencies of the group and the strong selection for more and smaller seeds, any slight change which would cause the pollen to clump together and be transferred in larger masses would be favored. It would be very easy for the insect to get stigmatic fluid on its back and cause larger masses of pollen to stick. Any variant that enhanced this tendency would be adaptive. In most cases, only the median anther would function in this way, and it is easy to see why the lateral anthers were lost in most lines of orchid evolution. As soon as the function of stigmatic fluid as adhesive had started, selection would favor the downward projection of the forward edge or lobe of the stigma, in order to more effectively coat the pollinator with glue. With this, we already have a rostellum, and the evolution of the orchids is well under way. The rest of the characteristic orchid features—compact pollinia, viscidia, caudicles, and stipes—are all improvements on the basic system, and we should not be surprised by their repeated evolution. At these early stages in orchid evolution, however, there were still several possibilities in the evolution of the relationship between rostellum and anther.

Vermeulen (1959) has drawn our attention to the rostellum-like projections from the lateral stigma lobes of *Epipactis gigantea*. One can imagine that a not-too-distant ancestor of *Epipactis* still had three functional anthers on a rather short column and, with a sack-like lip, all three anthers could still function in pollination. In one group the lateral anthers became more important, and selection shaped the lip into a one-way trap flower—in the front, out the side. This group went on to become the lady slippers. In most of the monandrous lines, the lateral anthers probably had been lost already, as they were also in the ancestor of *Epipactis* (which may not have been an ancestor of *Cypripedium*, but they were closely related).

We know nothing of the pollination of the Apostasioideae, except that some are self-pollinated. They do not seem to be gullet flowers, and one must conclude that they are not, by any means, ancestral orchids but rather a tangential line of evolution that did not amount to much.

In the spiranthoid line of evolution, the anthers did not extend far beyond the edge of the stigma, and the rostellum developed in relation with the apex of the anther and pollinia (acrotony). In such a case, selection favoring a viscidium and aggregation of the pollen would be strong. A well-developed viscidium is present throughout the spiranthoid lines. Further, a terminal viscidium is not so well suited to a gullet flower, and the Spiranthoideae are, as a group, primarily keyhole flowers.

In the primitive Orchidoideae, such as *Cephalanthera*, *Epipactis*, or *Chloraea*, the anther projects well beyond the stigma, and such a system can function with minimal development of rostellum and minimal aggregation of the pollen. Thus, we still find primitive stages of this line in the modern flora. The northern line, the Neottieae, are relictual, though the Listerinae have developed an interesting variant of the rostellum—the glue-drop under pressure. It is noteworthy that this group is still associated with wasps, unspecialized bees, and primitive flies (van der Pijl and Dodson, 1966; Ackerman and Mesler, 1979). Of the southern branch, the African element very early developed sectile pollinia and two viscidia in contact with the base of the anther. This condition gave rise to the incredible floral diversity of the Orchideae and Diseae. We suspect that the primitive Diseae did not have any spur, as spurs have since developed in nearly every perianth member of different groups. It is unlikely that so many different spurs would have developed if the group had started with any sort of spur. Of the remainder of the southern Orchidoideae, the South American element remained rather primitive, and the Australian element radiated greatly, with the relationship between rostellum and anther usually being somewhere between extreme basitony and extreme acrotony. Again, this system works well with loosely aggregated pollen, and that is what we find in many Diurideae.

In the Epidendroideae we find a distinct pattern of evolution. An early ancestor of the group evolved the "incumbent" anther, in which both the anther and the median stigma lobe bent downward toward the lip. This was clearly adaptive, as it assured that the departing pollinator would touch both the stigma and the anther. The mechanism was quickly perfected, probably then being something much like the modern *Vanilla* flower. This mechanism works fairly well, even with soft pollinia, and we thus have a nice series of primitive epidendroid orchids still with us. The mechanism works so well that there is less pressure for the development of a viscidium, and the viscidium has been developed repeatedly, but rather late within this line of evolution. In the advanced groups with a viscidium, of course, the originally incumbent position of the anther may be lost.

This leaves the Vandoideae, a rather isolated group whose relationships with other subfamilies are not altogether clear. In this line, I believe, the anther projected beyond the stigma and the rostellum developed in relation to the basal portion of the pollinia. In this case, however, the terminal portion of the anther sacks was suppressed, so that a short, operculate anther, superficially very like that of most Epidendroideae, evolved. Here again there was, in the very early stages, a minimally efficient system, so that selection for improvement was strong. Thus we find viscidia, and usually stipes, throughout the Vandoideae, and no clear links between this and other subfamilies.

Parallelism and Convergence

Once the primitive orchids had lost the three ventral (abaxial) anthers, these plants were already predisposed to a certain pattern of evolution, and many aspects of this pattern have occurred independently in several or many different groups of orchids. The rosette habit, for example, has evolved independently in several groups of terrestrials. Saprophytism crops up repeatedly, but this is scarcely surprising since all orchids go through a saprophytic stage. Epiphytism has evolved independently within the Cypripedioideae, the Spiranthoideae, and the Orchidoideae, as well as in the Vandoideae and Epidendroideae. Indications are that this habit evolved independently several times within each of these two groups, with, of course, some "reversions" to a terrestrial habit. The evolution of corms or pseudobulbs has occurred more than once, and the evolution from corm to pseudobulb of several internodes and on to pseudobulb of a single internode has occurred a number of times. Similarly, the evolution of conduplicate from plicate leaves has occurred in every subfamily and probably several times in some of them. The reduction of the seed to its typical form has occurred independently in at least three subfamilies, and the aggregation of the pollen to pollinia has occurred independently in at least two or three subfamilies. Sectile pollinia have evolved independently in at least three different groups (and probably twice within the Orchidoideae). Compact soft pollinia have evolved independently in the Cypripedioideae (in *Phragmipedium*), and the evolution of compact, more or less hard pollinia has occurred in the Diurideae, the Cranichideae, the Epidendroideae, and probably independently within the Vandoideae. The viscidium has evolved at least once in every monandrous subfamily and a number of times within the Epidendroideae. Brieger (in Schlechter, 1970) considers the presence of a viscidium to be primitive, but what selection pressure could account for the repeated loss of this efficient system? The stipe has evolved independently in the Prasophyllinae, in some Bulbophyllinae, in the Sunipiinae, and in the Calypsoeae, and I see no reason to believe that it has not evolved independently in different groups of the Vandoideae. This list of parallelisms should make one skeptical of any simple, one-feature classification of the Orchidaceae, or of any other family. One may, of course, conclude that my classification is at fault, but if we assume that any one of these features has arisen once and only once, then we make the other features even more strikingly polyphyletic.

The independent evolution of single, similar features, or parallelisms, should not cause too much trouble for the alert systematist, but convergence can be more deceptive. In convergence we find that adaptation to a similar habitat or life-style has led to parallelism in several different features. In the extreme case, this can lead to remarkably similar organisms that are only distantly related to each other. We find a number of convergences in growth habit and flower structure between *Bulbophyllum* and *Pleurothallis*, which would seem to be ecological analogs in the Old World and New World tropics. Similarly, *Jacquiniella* (Laeliinae, New World), *Cryptocentrum* (Maxillariinae, New World), and *Sepalosiphon* (Glomerinae, Asia) all have strikingly similar flowers, all probably adapted to moth pollination. With so much parallelism, we might expect some of the most highly evolved groups to converge, and this is indeed the case. If the monopodial *Pterostemma* (Oncidiinae) had been found in Asia, it would probably be classified in the Sarcanthinae without a qualm. Garay (Dunsterville and Garay, 1972) considers *Dunstervillea* to be a member of the Sarcanthinae; I believe it a member of the Ornithocephalinae. At persent we cannot be sure who is right. Once we know enough about pollen structure, anatomy, chromosome number, and embryology of this and other orchids, the problem should vanish.

Convergence between distantly related organisms is still not that big a problem. It is convergence between closely related species or genera that fools even the experts. For many years we used the names *Angraecum falcatum* and *A. philippinense*, and spoke of the curious geographic disjunction. Now we classify both *Neofinetia falcata* and *Amesiella philippinensis* in the Sarcanthinae, in spite of the strong floral resemblance to *Angraecum*, in each case doubtless imposed by adaptation to moth pollination. In the subtribe Oncidiinae we find an unusually good example of convergence that is only recently being recognized as such. There is strong evidence that several related groups have independently adapted to pollination by large anthophorid bees (usually *Centris*). Because these groups have similar flower structure, all have been classified as *Oncidium*. Anatomy, chromosome numbers, and compatibility all show that some of these groups are more closely related to other genera than to the bulk of the species known as *Oncidium*. Unfortunately, the type species of the genus *Oncidium* belongs to one of the most distinctive groups, which is more closely related to the *Comparettia* complex than to the rest of "*Oncidium*." Thus, purely nomenclature matters complicate the picture, despite the clear biological facts (Williams and Dressler, in prep.).

Overall Patterns

Current evidence suggests that an important center of angiosperm evolution was West Gondwanaland, that is, the continent that later split to form South America and Africa. Many authors have considered Southeast Asia and Australasia to be the cradle of flowering plant evolution, because of the primitive angiosperms that are to be found there now. However, this region simply did not exist until Miocene time, when the Australian plate approached Asia (Raven and Axelrod, 1974). We really cannot say very much about where the orchids first evolved, though a tropical area seems indicated. At present, the most primitive genera are either pantropical or well scattered. *Cypripedium* and the Neottieae are northern; *Palmorchis*, *Epistephium*, and *Selenipedium* are all South American, and *Diceratostele* is African. The Malaysian Apostasioideae are primitive but by no means ancestral.

Palmorchis is a plant of wet forests, as *Diceratostele* may also be, but most primitive orchids are plants of dryish forest or more open habitats. *Epistephium* and *Eriaxis*, especially, are genera of open savanna or sterile, rocky soil. It is quite likely that the first orchids were plants of such rocky sites, rather than forest plants. One finds, in fact, that spiral leaf arrangement is more typical of open habitats than of forests, where a distichous arrangement is more frequent. One may find this difference to occur within populations, as in the case of some *Cypripedium* species (Atwood, pers. comm.).

We now have a good deal of evidence on the radiation of the angiosperms during the Cretaceous period, with no firm evidence that they existed before that time (Hickey and Doyle, 1977). We may safely assume that the orchids did not evolve until after there were flower-visiting Hymenoptera, but the early fossil record of the aculeate Hymenoptera is only a little better than that of the orchids. There is, however, a fossil ant, *Sphecomyrmex*, that is known from the lower part of the upper Cretaceous period (Wilson, Carpenter, and Brown, 1967). As the ants were surely derived from wasps, we may feel sure that wasps were present during the upper Cretaceous (possibly before). For the Eocene epoch the insects have a rather good fossil record (due to amber deposits), and we find that at least five modern bee families were then present (Zeuner and Manning, 1976). Crepet (1979) has reviewed the fossil record of flowers and inflorescences and finds evidence of bee-pollinated flowers in the mid Eocene, as we would expect, from the bee's fossil record. He concludes, on the basis of the fossil record, that the orchids had not yet evolved then; but as far as the fossil record goes, there is little evidence that the orchids ever did evolve (Schmid and Schmid, 1977). I would suggest that orchids were certainly evolving by that time, and probably earlier. The distribution of the primitive orchids, such as *Vanilla*, *Tropidia*, and *Corymborkis*, and the distribution of the major groups of orchids points to the differentiation of the major lines of orchid evolution (the subfamilies) in the late Cretaceous or the very early Tertiary period, when South America and Africa were much closer together. Clearly, a major part of orchid evolution has taken place after the Paleocene epoch (more or less 55 million years ago, give or take a few million), while the three major tropical areas have been well isolated. There have been, of course, instances of long-distance dispersal since that time. I would judge that about 50 to 55 million years ago all six subfamilies were present in recognizable form, and primitive representatives of most tribes were on the scene then or relatively soon thereafter. In the early Tertiary period, the epidendroid and vandoid orchids were probably already occupying rock outcrops and other well-drained sites, and the evolution of epiphytes must have been going on through most of the Tertiary. Dispersal of orchids between Eurasia and North America was relatively easy up into the mid Tertiary, but the genera that show disjunctions between eastern Asia and southeastern North America suggest that this was a warm temperate corridor, but not really tropical.

We may ask what there is about the orchids that might explain their patterns of evolution. There are several factors to be considered (van der Pijl and Dodson, 1966). First, the orchids had rather fleshy roots with a velamen, which predisposed them for adaptation to well-drained sites, and eventually for epiphytism. Coupled with this was the tendency to saprophytism and the correlated tiny seeds. With this and a predisposition to insect pollination, it was very nearly inevitable that the orchids should take to the trees, and we have seen that there are at least a few epiphytes in five of the six subfamilies. Going a little further back, the relationship of the stigma and the median anther, coupled with the saprophytic tendencies, predisposed this group to its path of floral evolution. Once bilateral symmetry had become established, the whole pattern of evolution was already well started.

I visualize the first steps toward epiphytism as occurring in a seasonally dry climate (and most tropical climates are seasonally dry), and especially on cliffs and in rocky areas. Some orchids adapted to these habitats by developing corms, others by developing fleshy roots or leaves. With this pattern of evolution, then, the orchids were predisposed to adapt to an epiphytic life, perhaps first as humus epiphytes, but later diversifying and occupying many different epiphytic niches.

Classification

6

Many articles and books have been written on the philosophy and process of classification (too many, I think). Among the best are Mayr (1969), Simpson (1961), and Davis and Heywood (1963). Most authors agree that classification should be based on overall similarity. That is, two species or groups of species which resemble each other in many features are thought to be closely related, those that are very dissimilar are less closely related. By overall similarity, I do not mean superficial similarity. We can find many cases of plants that look very similar at a glance, as a result of parallelism or convergence, but prove to be very different in the details. We like to think that our genera, subtribes, tribes, and subfamilies all represent "natural," or phylogenetic, groups, and this is often given as a criterion for establishing taxonomic groups. However, we can rarely prove that suprageneric groups are natural, though we can eliminate many classifications as clearly unnatural. As we learn more about the groups we study, we can make better judgments of their phylogenetic significance. If a group is coherent, with lots of reticulate relationships within the group, and we can find no correlation of features which will clearly divide it into two, it is probably natural.

In recent years three main schools of thought have developed among systematic biologists, and they have generated a great deal of heated discussion, with the proponents of each philosophy often maintaining that theirs is the only truly scientific (or phylogenetic) viewpoint (see, for example, Mayr, 1965, 1974; Hennig, 1965; Bremer and Wanntorp, 1978). These schools of thought are:

Phenetic (or numerical) systematics. The pheneticists hold that classification should be based on overall similarity, and that all features should be given equal value. In some cases they maintain that classification should not be based on hypotheses of phylogeny but strictly on the features of the organisms to be classified. In practice, the pheneticist tabulates as many discrete features as possible for all the plants or groups under study, feeds that data into a computer with the appropriate program, and lets the machine do the classification. The main problems with this approach are that it is extremely expensive in both time and money, and that the results depend very much on the kinds of data gathered and the program used. These systems are useful, though, and one of their best aspects is that they force the student to consider many attributes of the plants carefully and systematically. In many groups of plants, this alone will help a great deal, even if one cannot find the funds to run the data through a computer. Phenetic studies of orchids have been published by Hall (1965), Wirth, Estabrook, and Rogers (1966), and Lavarack (1976).

Cladistic systematics. The cladists, at the other extreme, hold that phylogeny is the only possible basis for classification, and that phylogeny, and thus classification, can be determined by an analysis of the features of the organisms under study. The cladists draw a sharp distinction between ancestral (plesiomorph) and derived (apomorph) features, and hold that only derived features are of value in determining phylogeny or relationship. The cladists assume that all speciation occurs by dichotomy, and analyze the features of a group so as to determine the minimum number and sequence of branchings (speciations) which can have given rise to a group. Such analysis can be very valuable, and it is unfortunate that botanists have paid little attention to the cladist's methodology. (The cladists' abstruse terminology does little to help their cause.) The main weaknesses of the cladistic viewpoint are: (1) Their insistence on strict dichotomous branching. There are many patterns of speciation in nature, and some species may exist, as such, for long periods, while many peripheral species arise from a single parent species. (2) They ignore the degree of divergence among groups, and insist that all "sister groups" must be given the same rank. (3) For cladistic analysis to be meaningful, one must be able to distinguish between "uniquely derived features" and features that have evolved independently in related lines. Parallelism in related lines is a very real phenomenon and must be considered in any analysis.

Synthetic systematics. There remain a number of biologists who see something of value in each of the above philosophies. The adherents of this viewpoint try to consider all available data and find the best system of classification that fits the data. Organic diversity is so great that no simplistic formula will give consistently good results in its analysis.

"A priori key character method" Many orchid workers, especially, tend to choose a given character, such as number of pollinia or position

of inflorescence, and give this feature, *a priori*, great importance, regard-less of other features. This has been bad biology for two centuries, so much so that recent authors such as Mayr (1969) and Simpson (1961) discuss this unscientific approach only briefly. It is, of course, not a classification of plants but a classification of key features. Some authors seem to feel that there *should* be some one feature that will automatically distinguish genera or subtribes, wherever it occurs. Others, perhaps, feel that all classification is artificial, and why should they worry if their classifications are more artificial than others? This may have been the viewpoint of Pfitzer, when he chose easily seen vegetative features over floral details that might even be missing in a critical specimen. For a good example of this bad approach, see the characterization of the general *Auliza* and *Pleuranthium* in the new edition of Schlechter (1976). All *Epidendrum* species with lateral inflorescence are placed in the genus *Pleuranthium,* and all species with pseudobulbs are placed in *Auliza* (species which have both features are, for some reason, placed in *Auliza*). In each case, a glance at the illustrations shows that each "genus" is a hodge-podge of unrelated species (or species groups), each of which is closely related to species of *Epidendrum* with slender stems and terminal inflorescences. There is only one sin greater than the use of *a priori* key characters, and that is the use of *a priori* key characters without carefully observing the characters chosen. Examples of this, too, are to be found in the new edition of Schlechter, as in the *Gattungsreihe* Hexisieae. In this supposed group, only two genera actually show the features on which the group is supposedly based, and these two are not at all closely related to each other (Dressler, 1979a).

Connectedness. A useful approach to classification is to emphasize the relationships, or coherence, within a group. We may find close relation-ships (or high overall similarity, if you prefer) between species A and B, between B and C, C and D, and so on, until we have species A connected to the very different species Z. *After* we have grouped all these species together, we can search for some neat key features that will permit us to easily identify the group as such. If we cannot find an easy key feature, it does not necessarily mean that the group is not a natural one. It may only mean that we must write a more complicated key.

CLASSIFICATION VERSUS IDENTIFICATION

Some classification problems, especially in botany, seem to be caused by a confusion of classification and identification. These two processes are clearly related, but they are not the same thing. To use an extreme example, frogs and tadpoles are drastically different, yet we classify them as ontogenetic phases of the same species. We cannot identify them by the same key features; indeed, we must make separate keys for frogs and tadpoles, but we do not put them in different taxonomic categories. Similarly, *Phyllanthus fluitans* looks more like the water fern *Salvinia* than it does like an euphorbiaceous shrub, yet the structure of the flowers and seeds show it to be a member of the euphorbiaceous genus *Phyllanthus*. For purposes of identification, we may separate water plants from land plants, epiphytes from terrestrials, herbs from trees, blue flowers from yellow flowers, or large flowers from small ones, but we do not base formal taxonomic categories on these distinctions. Some botanists have actually suggested that the parasitic genus *Lathraea* should be placed in the family Orobanchaceae, rather than in the Scrophulariaceae, where studies have shown it to belong, as this makes it simpler to key (Davis, 1978). To my way of thinking, this is a confusion of classification and identification. We do not place *Wolfia* and *Wolfiella* in the green algae, though I am sure that would make most specimens easier to key out. Perhaps our great reliance on dichotomous keys in identification causes people to expect taxa to sort out neatly on simple features. When we are lucky, we may be able to separate species, and even genera, by fairly simple keys, but subfamilies and families are synthetic units and should include diverse phyletic trends.

SPLITTERS AND LUMPERS

We often hear the comment that some botanist is a "splitter," or a "lumper," and the members of each category tend to disparage those of the other. The extreme lumper will hold his specimens (dry specimens, of course) at arm's length, and if he doesn't immediately see a difference, he will decide that they are all the same species. The extreme splitter, on the other hand, will usually give specific status to every name that has been published, on the theory that they *may* be different (after all, they have been named). There is also that curious creature, the "splimper," who sees important differences in the group that he is studying, and so he recognizes many genera and species in his own group, while preferring to lump the species and genera of other groups with which he is less familiar. Either splitting or lumping, when they are taken as general policy, will lead to a poor classification. One practical criterion for a good taxonomic study is that the author will both split and lump within the same group, depending on the features of the plants and their distinction and variation.

Criteria for Relationship

What, then, are the criteria that one should use for selecting key features? They are very simple. One should consider all features of a group in delimiting the group as such. Once the group is delimited, one may choose any key character that works as an aid in the recognition of the group. The point is that a feature that works beautifully on one group will fail miserably in another. Further, nearly all large genera have a few glaring exceptions to any one key feature that is chosen (such as the few *Epidendrum* and *Pleurothallis* species with lateral inflorescences). This does not necessarily mean that the exceptions do not belong in the genera in question. Only if one finds other differences that are correlated with the divergent features may there be some cause for revision.

Even though all features of a plant, whether visible or not, are criteria for relationship, there are some special classes of criteria that call for a few comments. In theory, the ideal data for relationships and phylogeny would come from fossils. We have found no fossil orchids, and are unlikely ever to find many. We might someday find some fossil pollen that would give us a few hints about the phylogeny of the orchids, but a really good fossil record is quite unlikely. (For a detailed review of all supposed orchid fossils, see Schmid and Schmid, 1977.)

Floral versus vegetative features. Most taxonomists give primary importance to floral features and feel that vegetative features are so evolutionarily plastic that they are less dependable. I have also heard the exact opposite viewpoint defended. In any case, flowers, especially orchid flowers, are complicated mechanisms with many features. While size and color are doubtless plastic, the basic structure of the flower and the column are not easily changed to a different type. One cannot imagine an *Epidendrum* simulating the floral details of a *Vanda* or a *Spiranthes*. These flowers differ not just in size, shape, and color, but in their whole organization.

Organizational features. This refers to "fundamental" features involving the relationship between different parts of the organism. They surely represent the action of many genes, rather than only a few, and systematists usually assign great significance to these features, whatever their rationale. Perhaps the best examples of this sort of feature are to be found in the animal world; we may mention internal versus external skeleton, or the type of skin (smooth or with scales, feathers, or hair), both of which are important in animal classification. With reference to plants, we may list scattered vascular bundles versus a ring of vascular bundles (monocot versus dicot), bilateral symmetry versus radial symmetry in the flower, the incumbent anther of the epidendroid orchids, or stipes and caudicles, as such features. Some of these features may have evolved several different times, and some of them may be drastically modified by a single mutation, but we still consider them to be much more important than simpler differences of size, form, color, and so on.

A related concept is that of "neutral" features. In theory, at least, these features are not affected by selection, and they are thus better markers for phyletic groups than features subject to selection pressure. The presence of subsidiary cells, for example, may have some adaptive value, but the way in which these cells develop (which is rather an organizational feature) may be of little adaptive significance. For this reason, the developmental pattern of the epidermis may be much more meaningful in classification than more obvious features that probably have evolved independently in different groups.

Microscopic features. Anatomical details are often found to be of value in classification. If they have any special drawback, it is simply that they have not been well sampled and that the sampling is tedious. The details of both pollen structure and seed structure are very promising, and I hope that the sampling of these features will be greatly extended in the near future. The ontogeny of the flower can help in determining the homologies of the mature flower. In general, I believe that observation of the three-dimensional primordium with a stereoscopic dissecting microscope is much clearer than trying to recreate the structure from microscopic sections.

Biochemistry. Biochemical data have proven to be enormously useful in some plant groups. Again, the problem is one of sampling. So far, the data that we have on the biochemistry of orchids are very spotty.

Chromosome number. As microscopically visible carriers of genetic material, the chromosomes are usually considered to have special importance as indicators of relationship. There are many cases in which differences in chromosome number correspond to generic lines in the orchids, and other cases in which whole subtribes show little or no variation in chromosome number. The frustrating feature of the Orchidaceae is that one finds a diploid number of 38 or 40 in some members of nearly every subtribe throughout the family. When the patterns of evolution within the subtribes have been worked out, we may be able to see some overall patterns for the whole family (Jones, 1974).

Crossability. We may take the existence of a confirmed hybrid between two species as clear evidence of relationship. Further, by determining the fertility of the hybrid or the degree of chromosome pairing which it shows, we may get some measure of the degree of relationship.

As we have already noted, crossability is usually confined to genera within a subtribe, but as many orchidists continue to try unlikely crosses, we should not be surprised by a few intertribal hybrids. We would, however, expect them to be weak growers and quite sterile. Either natural or artificial intergeneric hybrids are known in some fourteen orchid groups.

Natural hybrids. We must draw a clear distinction between artificial and natural hybrids. Artificial hybrids may be produced between species which are strongly isolated by prepollination barriers and could never produce hybrids, much less exchange genes, under natural conditions. In most cases natural hybrids are restricted to intrageneric crosses and crosses between very closely allied genera. If we eliminate the natural hybrids between *Cattleya* and the section *Cattleyodes* of *Laelia* (a very artificial generic distinction) and many crosses between rather dubious genera of the European Orchideae, such as *Orchis* and *Aceras*, only a handful of natural intergeneric hybrids would remain.

While natural intergeneric hybrids usually involve closely related genera, we do have one apparent case of a natural intertribal hybrid. Plants which appear to be *Epipactis palustris* (Neottieae) × *Gymnadenia conopsea* (Orchideae) have been found in Italy. Specimens are deposited in the Kew Herbarium, and Peter Taylor is preparing a paper on these remarkable plants. As one would expect, the putative hybrids are sterile, or at least they produce no pollen.

How to Judge a Classification
We sometimes read that classification is purely a matter of opinion. This is clearly absurd, though at times such a view is even expressed by people who claim to be biologists. If we know nothing whatever about a group of plants, then we can pick their classification by Ouija board. As soon as we have any knowledge about the plants, we have some basis for choosing one classification over another. There are cases, of course, when classification *is* a matter of opinion. When there are two distinct but closely allied groups of species, they may be treated as subgenera of a single genus or as two closely allied genera, and neither arrangement does violence to the pattern of relationship. In the classification used here, on the basis of current knowledge the Cryptostylidinae may be treated as a distinct subtribe or an alliance of the Cranichidinae. Similarly, one may treat the Orchideae and the Diseae as two closely related tribes, or as two groups of subtribes within a single, natural tribe. In each case, further material or more details of

Plate 1.   Habitats

1. Cloud forest, Panama
2. **Swamp** with *Arethusa*, Pennsylvania, United States
3. High altitude vegetation (subparamo), Ecuador
4. Grassland with *Calopogon*, Florida, United States
5. Grassy roadside with *Arundina*, West Malaysia
6. *Encyclia adenocarpon* on *Pachycereus*, thorn forest, western Mexico

1

2

3, 4

5, 6

Plate 2. Habits terrestrial

7. *Corallorhiza striata* (Corallorhizinae), Michigan, United States
8. *Epipogium roseum* (Epipogieae), Australia
9. *Dactylorhiza aristata* (Orchidinae), Alaska
10. *Corymborkis veratrifolia* (Tropidiinae), Papua New Guinea
11. *Anoectochilus imitans* (Goodyerinae), New Caledonia
12. *Lyperanthus nigricans* (Caladeniinae), Australia

7   8   9

10

11, 12

Plate 3. Habits epiphytic

13. *Notylia linearis* (Oncidiinae), Panama
14. *Dendrobium crumenatum* (Dendrobiinae), Malaysia
15. *Bulbophyllum nematocaulon* (Bulbophyllinae), West Malaysia
16. *Cymbidium madidum* (Cyrtopodiinae), seedling on *Melaleuca*, Australia
17. *Vanda tricolor* (Sarcanthinae), Indonesia
18. *Lepanthes* sp. (near *calodictyon*) (Pleurothallidinae), Colombia

13

14

15

16, 17

18

Plate 4. Pollination

19. *Platanthera ciliaris. Papilio troilus* has a pollinarium near the base of its tongue.
20. *Coryanthes rodriguezii* with *Eufriesea superba.* The bee is emerging from the flower.
21. *Cryptostylis erecta* with *Lissopimpla excelsa.* The wasp has pollinia on its abdomen.
22. *Chiloglottis formicifera* with a male thynnid wasp.
23. *Chiloglottis formicifera.* The antlike callus gave this species its name, but the
    flower is actually mimicking a wingless female wasp.

19

20

21

22, 23

Plate 5. Apostasioideae, Cypripedioideae, Spiranthoideae

24. *Cypripedium acaule* (Cypripedioideae), northeastern United States
25. *Tropidia polystachya* (Tropidiinae), Florida, United States
26. *Stenorrhynchos speciosum* (Spiranthinae), Costa Rica
27. *Pristiglottis montana* (Goodyerinae), New Caledonia
28. *Ponthieva brenesii* (Cranichidinae), Panama
29. *Apostasia wallichii* (Apostasioideae), Australia

24   25   26

27, 28   29

Plate 6. Orchidoideae Neottieae, Diurideae

30. *Listera cordata* (Listerinae), Colorado, United States
31. *Chloraea gavilu* (Chloraeinae), Chile
32. *Epipactis gigantea* (Limodorinae), western United States
33. *Corybas fimbriatus* (Acianthinae), Australia
34. *Elythranthera emarginata* (Caladeniinae), Australia
35. *Calochilus campestris* (Diuridinae), Australia

30

31

32

33

34

35

Plate 7. Orchidoideae Orchideae, Diseae. Triphoreae

36. *Satyrium hallacki* subsp. *ocellatum* (Satyriinae), southern Africa
37. *Psilochilus carinatus* (Triphoreae), Panama
38. *Habenaria entomantha* (Habenariinae), Panama
39. *Disperis capensis* (Coryciinae), southern Africa
40. *Orchis militaris* (Orchidinae), Europe
41. *Brownleea caerulea* (Disinae), southern Africa

36   37   38

39, 40   41

Plate 8. Epidendroideae   Vanilleae, Gastrodieae. Wullschlaegelieae

42. *Vanilla barbellata* (Vanillinae), Florida, United States
43. *Wullschlaegelia calcarata* (Wullschlaegelieae), Panama
44. *Gastrodia sesamoides* (Gastrodiinae), Australia
45. *Cryptanthemis slateri* (Rhizanthellinae), Australia
46. *Eriaxis rigida* (Vanillinae), New Caledonia
47. *Cleistes rosea* (Pogoniinae), Panama

42   43   44

45, 46   47

Plate 9. Epidendroideae Arethuseae, Coelogyneae, Malaxideae

48. *Coelogyne radicosa* (Coelogyninae), West Malaysia
49. *Hexalectris spicata* (Bletiinae), Florida, United States
50. *Sobralia rosea* (Sobraliinae), Ecuador
51. *Arethusa bulbosa* (Arethusinae), northern United States
52. *Liparis lacerata* (Malaxideae), Malaysia
53. *Thunia alba* (Thuniinae), Asia

48

49

50, 51

52

53

Plate 10. Epidendroideae Epidendreae I

54. *Encyclia venosa* (Laeliinae), Mexico
55. *Eria hyacinthoides* (Eriinae), Malaysia
56. *Aglossorhyncha jabiensis* (Glomerinae), Papua New Guinea
57. *Glomera obtusa* (?) (Glomerinae), Papua New Guinea
58. *Mediocalcar abbreviatum* (?) (Eriinae), Papua New Guinea
59. *Epidendrum hunterianum* (Laeliinae), Panama

54

55

56, 57

58, 59

Plate 11. Epidendroideae Epidendreae II, Cryptarrheneae, Calypsoeae

60. *Cryptarrhena guatemalensis* (Cryptarrheneae), Panama
61. *Calypso bulbosa* (Calypsoeae), Michigan, United States
62. *Octarrhena condensata* (Thelasiinae), West Malaysia
63. *Dendrobium stratiotes* (Dendrobiinae), Indonesia
64. *Sunipia racemosa* (Sunipiinae), Thailand
65. *Dendrobium sophronites* (Dendrobiinae), Papua New Guinea

60 61 62

63, 64

65

Plate 12.   Epidendroideae   Epidendreae–Pleurothallidinae

66. *Platystele minimiflora*, Panama
67. *Lepanthes* sp., Panama
68. *Pleurothallis ignivomi*, Ecuador
69. *Barbosella orbicularis*, Panama
70. *Dracula vampira*, Ecuador
71. *Masdevallia angulifera*, Colombia

Plate 13. Epidendroideae Epidendreae–Bulbophyllinae

72. *Bulbophyllum globuliforme*, Australia
73. *Bulbophyllum subcubium* (?), Papua New Guinea
74. *Bulbophyllum (Hapalochilus) callipes* (?), Papua New Guinea
75. *Bulbophyllum rothschildianum*, tropical Asia
76. *Saccoglossum papuanum*, Papua New Guinea
77. *Pedilochilus flavum*, Papua New Guinea

72   73

74, 75   76, 77

Plate 14. Vandoideae Polystachyeae, Vandeae

78. *Polystachya bella* (Polystachyeae), Uganda
79. *Chiloschista lunifera* (Sarcanthinae), Thailand
80. *Vanda insignis* (Sarcanthinae), Indonesia
81. *Taeniophyllum trachypus* (Sarcanthinae), New Caledonia
82. *Rangaeris muscicola* (Aerangidinae), southern Africa
83. *Neofinetia falcata* (Sarcanthinae), Japan

78   79   80

81   82, 83

Plate 15. Vandoideae Maxillarieae

84. *Cochleanthes aromatica* (Zygopetalinae), Costa Rica
85. *Telipogon* sp. (Telipogoninae), Costa Rica
86. *Sphyrastylis escobariana* (Ornithocephalinae), Colombia
87. *Govenia purpusii* (Corallorhizinae), Mexico
88. *Dichaea panamensis* (Dichaeinae), Panama
89. *Maxillaria fulgens* (Maxillariinae), Panama

84   85

86, 87   88, 89

Plate 16.   Vandoideae   Cymbidieae

90. *Acriopsis javanica* (Acriopsidinae), Papua New Guinea
91. *Oncidium (Cyrtochilum) serratum* (Oncidiinae), Colombia
92. *Stanhopea costaricense* Stanhopeinae), Costa Rica
93. *Catasetum barbatum* (Catasetinae), South America
94. *Galeandra baueri* (Cyrtopodiinae), tropical America
95. *Polycycnis aurita* (Stanhopeinae), Colombia

90   91

92, 93   94, 95

flower or pollen structure or some other features might easily change the judgment that I have made here.

We must ask of an author how much evidence he offers, and how consistent it is with the known facts. A classification may be tested by studies of anatomy, cytology, and many other aspects of botany. It is this continual testing and checking that gives meaning to systematic botany. When we consider any classification, we may ask some simple questions: Does the classification separate distantly related (dissimilar) plants or plant groups? Does it place together closely related (similar) plants and groups? Is the system consistent? That is, do the features of the plants agree with what the classification says they should be? If the author offers a phylogenetic tree, or other evolutionary scheme, is this based on a careful analysis of features, or does it seem to be made from whole cloth, or based on preconceived ideas?

The Nomenclatural Hierarchy

In our consideration of classification, we will be concerned primarily with the categories between family and genus. Part of the hierarchy of categories established by the International Code of Botanical Nomenclature is given herewith.

Category	Ending	Examples
Order	–ales	Liliales
Family	–aceae	Orchidaceae, Liliaceae
Subfamily	–oideae	Orchidoideae, Epidendroideae
Tribe	–eae	Orchideae, Vandeae, Neottieae
Subtribe	–inae	Orchidinae, Laeliinae, Oncidiinae
Genus (plural genera)	Not fixed	*Orchis, Epidendrum, Vanda*
Species (singular and plural)	Not fixed, but must agree with gender of genus	*Orchis purpurea, Vanda tricolor, Epidendrum nocturnum*

Below the rank of species the code permits the use of subspecies, variety (or *varietas*), and form (or *forma*), in that order, but we will not be much concerned with these categories here. The categories outlined above are not really enough to show all the complexities of relationship and phylogeny, and there is often a tendency to add formal or informal categories, such as supertribe, alliance, subgenus, and so on. However, we find that there is no way in the world to show all the complexity on a piece of paper, so we will try to keep the system fairly simple.

Botanists have been rather lax about the nomenclature of tribes and subtribes, so that different authors may use different names for the same groups. In general, the simplest procedure is to follow the rules of nomenclature and use the correct name for each group. This is the easiest way to avoid confusion and promote communication. There is, however, one minor point on which I do not follow the latest rules of nomenclature, and this involves the spelling of names based on names ending in *er* (*Teuscheria, sanderiana,* for example). Since the last International Botanical Congress, the rules recommend that these names be spelled without an *i* (*Teuschera, sanderana*). This is a recommendation, rather than a requirement, and many of us prefer the traditional spellings.

History of Orchid Classification

The first real attempt to subdivide the orchid family was published by Lindley in 1826, when he recognized 8 tribes. He later found one of the distinctions difficult to use and reduced the number of tribes to 7 (see table 6.1), with "sections" or "divisions" delineated in some of them. Lindley was a first-rate botanist with a keen eye for relationships. His classification was a good one, and I am sure it would have been even better if he had had better microscopes and more living orchid material to study. Bentham modified Lindley's system for use in Genera Plantarum, and he recognized only 5 tribes, with 27 subtribes.

H. G. Reichenbach, Jr., an active orchid taxonomist for many years who made an important early contribution to the use of pollinia for classification (1852), wrote relatively little on the problem of subfamilies and tribes. He reviewed orchid classification in 1884 but without assigning any clear rank to the few subdivisions that he recognized within the family. Reichenbach had a talent for seeing relationships, could often catch the essence of a flower with a very simple sketch, and named many genera and species, but he was somewhat arbitrary and capricious, changing at times from one extreme to another (splitting and lumping). Unfortunately, he was more critical of others' work than he was of his own, and his descriptions were often too brief to carry much information. He later became irritated that some Englishmen had the nerve not only to work on orchids but to question Reichenbach's judgment on such matters. His ire was so great that he specified in his will that his herbarium should remain locked up and unavailable for study for twenty-five years after his death.

Table 6.1 A comparison of some major systems of orchid classification.

Lindley	Schlechter	Garay	Dressler
Family Apostasiaceae	Family Apostasiaceae	Family Orchidaceae	Family Orchidaceae
		Subfamily Apostasioideae	Subfamily Apostasioideae
Family Orchidaceae	Family Orchidaceae		
Tribe Cypripedieae	Subfamily Diandrae	Subfamily Cypripedioideae	Subfamily Cypripedioideae
Tribe Ophrydeae	Subfamily Monandrae	Subfamily Orchidoideae	Subfamily Orchidoideae
	Tribe Ophrydeae	Tribe Orchideae	Tribe Orchideae
			Tribe Diseae
Tribe Neottieae		Subfamily Neottioideae	
		Tribe Neottieae	Tribe Neottieae
	Tribe Polychondreae		Tribe Diurideae
		Tribe Cranichideae	Subfamily Spiranthoideae
			Subfamily Epidendroideae
		Tribe Epipogieae	Tribe Epipogieae
			Tribe Vanilleae
			Tribe Gastrodieae
Tribe Arethuseae			Tribe Arethuseae
Tribe Epidendreae	Tribe Kerosphaereae	Subfamily Epidendroideae	Tribe Coelogyneae
Tribe Malaxideae		Tribe Epidendreae	Tribe Epidendreae
			Tribe Malaxideae
Tribe Vandeae		Tribe Vandeae	Subfamily Vandoideae
			Tribe Polystachyeae
			Tribe Vandeae
			Tribe Maxillarieae
			Tribe Cymbidieae

Ernst Pfitzer criticized the system of Bentham in 1887 and offered a new system based primarily on easily seen vegetative features. Pfitzer delineated 32 tribes, with a number of subtribes. Pfitzer's classification was highly artificial, though the prize for artificiality must surely go to Beer (1863), who stoutly defended a classification of all orchids into 5 "Sippen," which were based on flower shape. Thus, *Angraecum*, *Habenaria*, and *Cypripedium*, for example, were grouped together. Pfitzer was a reasonably competent taxonomist and morphologist, in spite of his strange ideas on tribal and subtribal classification. His contemporary, Fritz Kränzlin, however, epitomized the worst of German taxonomy of that era. Kränzlin published a number of "revisions" and many new species. Unfortunately, he did not have an eye for relationships and his work was uniformly bad. A number of his new species were described not only in the wrong genera but in the wrong tribes or subfamilies.

Rudolph Schlechter probably described more orchid species than anyone else before or since. His descriptions are reasonably detailed, but his herbarium was destroyed during World War II, thus causing even greater problems than the temporary embargo of Reichenbach's herbarium. Schlechter tended to be a splitter, and seemed to feel that orchids in different countries should be different species. I believe that his work in Africa and New Guinea, where he had some field experience, was better than his work with American orchids. Schlechter wrote a revised classification of the family, which was published after his death (1926). Schlechter recognized only 4 tribes, and had 80 subtribes. While he eliminated some of the inconsistencies in Pfitzer's system, the names he used were based on Pfitzer's system rather than on the earlier work of Lindley and Bentham, whose names have priority under the rules of nomenclature. Schlechter's system was dichotomous and curiously asymmetrical, a feature which he may have felt to be scientific (fig. 6.1). Worked out in detail, with a key to subtribes and a

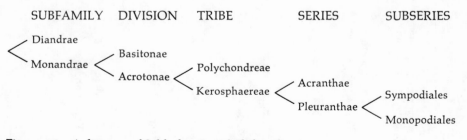

Figure 6.1 A diagram of Schlechter's orchid classification.

complete list of genera, Schlechter's system became almost the only system in use for a few decades. Mansfeld (1937a, 1937b, 1954) made some improvements on Schlechter's system, but other authors have seemed unaware of them. In 1960 Dodson and I did a review of orchid classification and made some modifications of Schlechter's system, largely to bring it into line with the rules of nomenclature (Dressler and Dodson, 1960). At about the same time Garay suggested the division of the family into 5 subfamilies. Recent authors have generally accepted the division of the family into 3 to 6 subfamilies but without much agreement on how to delineate the subfamilies. Working with names above the rank of genus is always a problem, and Butzin (1971) has done a great service for orchid classification by bringing together the names that have been used for subfamilies, tribes, and subtribes.

I do not want to leave the impression that all orchid taxonomists have been poor botanists. Workers such as Holttum, Mansfeld, Rolfe, J. J. Smith, and Summerhayes were all careful taxonomists who have done very good work. For the most part, however, they concentrated on problems at the specific and generic level and, except for Mansfeld, did not have much to say about subfamilies or tribes.

Classification To Be Used Here

The system which I will use in this book is outlined below, and diagrammed in figures 6.2 and 6.3.

Subfamily Apostasioideae
Subfamily Cypripedioideae
Subfamily Spiranthoideae
    Tribe Erythrodeae
    Tribe Cranichideae
Subfamily Orchidoideae
    Tribe Neottieae
    Tribe Diurideae
    Tribe Orchideae
    Tribe Diseae

Anomalous tribes
    Tribe Triphoreae
    Tribe Wullschlaegelieae

Subfamily Epidendroideae
    Tribe Vanilleae

Tribe Gastrodieae
Tribe Epipogieae
Tribe Arethuseae
Tribe Coelogyneae
Tribe Malaxideae
Tribe Cryptarrheneae
Tribe Calypsoeae
Tribe Epidendreae
Subfamily Vandoideae
Tribe Polystachyeae
Tribe Vandeae
Tribe Maxillarieae
Tribe Cymbidieae

The main feature in which this classification differs from other recent systems is in the treatment of the orchids with soft, more or less mealy pollinia. Garay (1960) essentially raised the *Polychondreae* of Schlechter to subfamily status as the Neottioideae, and several other authors have used the subfamily in much the same way. Others have followed Mansfeld in assigning some or all of the genera with incumbent anthers to the Epidendroideae, but this still leaves a rather diverse lot of orchids. Lavarack, in his extremely useful study (1971), has clearly indicated that several diverse lines were being lumped together as the Neottioideae. Actually it is not unusual for botanists to lump together the more primitive members of a group during early attempts at a classification. A little thought will show, though, that the common possession of a single primitive, or ancestral, feature such as soft, mealy pollen is scarcely a dependable indicator of relationship. Further, in rather oversimplified terms, the primitive elements of any family have had the longest time for evolution, and they are very likely to represent several diverse lines of evolutionary development. This is certainly the case with the orchids (compare figs. 6.2, 6.3, 6.4).

To separate the orchids with soft pollinia and incumbent anthers from the rest of the epidendroid line of evolution is arbitrary, as there is no sharp line at all, and is clearly cutting across one of the major lines of evolution. Accordingly, I assign the Vanilleae, the Gastrodieae, the Epipogieae, and the Arethuseae to the Epidendroideae, where they fit very well as less specialized members of that line of evolution.

In the remaining groups with soft pollinia, the Spiranthoideae form a coherent group, easily separated from the rest by their column structure and by the occurrence of mesoperigenous subsidiary cells associated

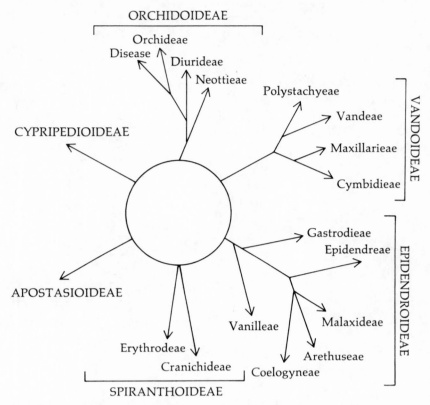

ORCHIDOIDEAE

Orchideae
Disease
Diurideae
Neottieae

CYPRIPEDIOIDEAE

Polystachyeae
Vandeae
Maxillarieae
Cymbidieae

VANDOIDEAE

APOSTASIOIDEAE

Gastrodieae
Epidendreae
Malaxideae
Vanilleae
Arethuseae
Erythrodeae
Cranichideae
Coelogyneae

EPIDENDROIDEAE

SPIRANTHOIDEAE

Figure 6.2 A diagram of the probable relationships of the major tribes of Orchidaceae.

with the stomata (Williams, 1975). There are no clear links between the Erythrodeae and the Cranichideae, but the similarities in flower structure suggest that they should remain together.

The remaining groups with soft pollinia lack subsidiary cells altogether and virtually always have the anther projecting well beyond the stigma. Most classifications exaggerate the isolation of the tribes Orchideae and Diseae, and Vermeulen (1966) even separates all monandrous orchids into two subfamilies, the Orchideae and the Diseae, as the Orchidoideae, and everything else as the Epidendroideae. This is very close to "splimping." Actually, there are a great number of resemblances between the Orchideae and the Diurideae. The presence of the unique root-stem tuberoid is, by itself, compelling evidence of a relationship. Basitony is found among the Diurideae, though the basically double viscidium of the Orchideae and Diseae is not. Barthlott

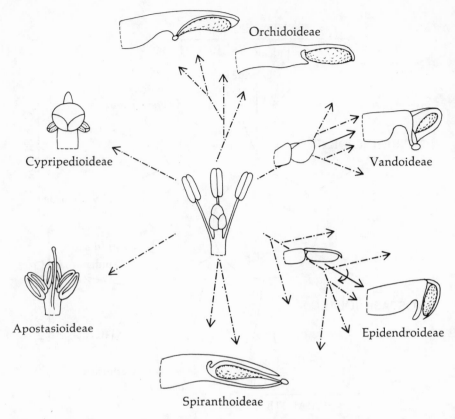

Figure 6.3 A diagram contrasting the subfamilies of Orchidaceae.

(1976b) finds seed structure similar to that of the Diurideae in *Satyrium* and *Cynorkis*. Placing the Orchideae, Diseae, and Diurideae, together with the Neottieae, in a single subfamily creates a natural phyletic group. The natural hybrid between *Epipactis* and *Gymnadenia* must also be considered as strong evidence of a relationship between the Neottieae and the Orchideae.

Of the orchid tribes with predominantly hard pollinia, I was, at one time, skeptical of the segregation into two subfamilies, Epindendroideae and Vandoideae, until I realized the implications of Hirmer's study of anther development. As closely as these two groups parallel each other, they are separate phyletic groups and have developed similar column structure in different ways, though they may be closely related.

Aside from differences in the treatment of orchids with soft pollinia, the classification used here stresses the details of the anther and pollinia, as have most other classifications of this family. Ideally, one should

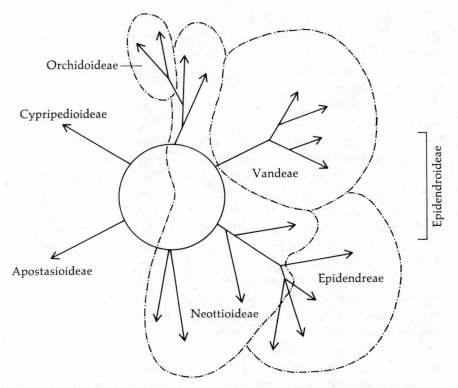

Figure 6.4 A diagram contrasting the classification of Garay with that used here. The subfamilies of Garay (dashed lines) are superimposed on the framework of figure 6.2. Garay's classification separates the Orchideae and Diseae (as a subfamily) from their closest allies, and places part or all of three different phyletic groups in the subfamily Neottioideae.

study fresh material of all orchid groups, for published descriptions are not always dependable or sufficiently detailed, and dried specimens are not satisfactory, especially in the groups with soft pollinia. I feel that this classification divides the orchids into phyletic groups in a more natural way than have previous attempts, and I have clarified some groups at the tribal and subtribal levels. However, question marks are still sprinkled liberally through the system. Where should we place the Triphoreae and *Wullschlaegelia*? Do the Tropidiinae really belong with the Spiranthoideae, or should they be placed in limbo, with the anomalous tribes, or elsewhere? The Bletiinae, the Corallorhizinae, and especially the Cyrtopodiinae are each rather diverse, and it is possible that some of them should be subdivided, but it is not clear what this division should be. Possibly the *Cranichis* alliance and the *Prescottia* alliance should be separated. The Glomerinae are now more clearly delineated, but what are their closest allies? These are some of the major questions, at least in my own mind. I will try to indicate these and other areas of uncertainty under the appropriate groups in the following chapters.

"Systematic" Order

In my classification, the groups are arranged in something approximating the traditional order, with more primitive groups at the beginning and advanced and specialized groups at the end. We find, though, that a linear arrangement imposes severe limits on any attempt to show phylogeny. We place the Apostasioideae first, as they are the least specialized. The Cypripedioideae follow, as they, like the apostasioids, have fertile lateral anthers, but there is no close relationship between these two groups. The Spiranthoideae and the Orchidoideae follow, and I place the Spiranthoideae first, because they are relatively isolated and less specialized than the highly evolved Orchideae and Diseae (but more so than the Neottieae). In most features, the Epidendroideae seem less specialized than the Vandoideae, and so the Vandoideae are placed last. Within the Vandoideae, I have placed the Cymbidieae last, as the reduction from four to two pollinia would seem to represent a further specialization, but, again, any attempt to arrange these groups in a linear order has its drawbacks.

A diagram such as figures 6.2 and 6.3, which shows all major groups diverging from a common source, is more satisfactory. Here, we can place the Apostasioideae and the Spiranthoideae next to each other, as there may be some relationship there, though both groups are quite isolated from all other orchids. At the same time, the Cypripedioideae and the Orchidoideae are adjacent, and there clearly seems to be close relationship between these two lines. In this diagram, I have placed the Vandoideae adjacent to the Orchidoideae, as they both have a similar anther position in their more primitive members, while the Epidendroideae show a basically different column structure. It does not matter much, though, which of these two subfamilies is placed closer to the Orchidoideae. The Cypripedioideae, Orchidoideae, Epidendroideae, and Vandoideae all seem to be interrelated.

The following chapters will take up the subfamilies, tribes, and subtribes in this "systematic" order. We will be concerned primarily with the overall classification of the family, so that we cannot go into detail at the generic level. I will list the genera, as best I can, for each subtribe (or larger unit, if not subdivided), but the project would get quite out of hand if I tried to do much more. The hobbyist's greatest interest will probably be in seeing which genera are placed in the same subtribe, to get some idea of the genera that may possibly be crossed with each other; I know of one grower who deliberately set out to test my earlier ideas on classification in this way. In each case, I will give as much general information on the group as I can, and I will refer to more detailed studies or reviews, if such are available. For each group I will give an estimate of the number of species. In some few cases this may be close to the actual number of species known. Where I have no special knowledge of the number of species, my estimate will be based on the most recent edition of Willis' *A Dictionary of the Flowering Plants and Ferns* (1973). This gives a relatively unbiased estimate, which is probably on the low side in most cases. The chromosome numbers given are diploid (2n) numbers, and I will not list tetraploid and higher numbers when that is clearly their status. The treatment of subtribes will, of necessity, be rather uneven. I know some groups much better than others, and for those groups that I do not know well, there is much more published information on some than on others.

In most cases, the names of subfamilies, tribes, and subtribes used here are among those listed by Butzin (1971), and his paper should be consulted for details of publication.

The Apostasioid and
Cypripedioid Orchids

7

Considering the enormous diversity and specialization of the orchid
family, it is a bit surprising to find two different groups with fertile
lateral anthers among the living orchids, the Apostasioideae and
Cypripedioideae. These two groups are not closely related to each
other, and neither was ancestral to any other group of living orchids.
Rather, these groups went off on evolutionary tangents that preserved
some primitive features. They have been evolving along these tangential
pathways for a long time, however, and they are not at all like the
common ancestors that each shares with some other orchid group.
Still, by studying these and other primitive living orchids, we may get
some idea of what their remote ancestors were like.

Subfamily

APOSTASIOIDEAE Reichb. f.

DESCRIPTION   Habit: forest floor terrestrial, with elongate stems; nodular
storage roots in *Apostasia*. Leaves: spiral, convolute, plicate, nonarticulate. Inflorescence: terminal, erect, or spreading, simple or branched,
flowers spiral. Flowers: small, white, yellowish, or yellow, perianth
with bilateral symmetry, resupinate or not, anthers 2 or 3, the median
anther, when sterile, may be represented by a fingerlike staminode;
filaments partly united with the style; pollen grains as monads, not
clumped; style slender, stigma lobes equal and similar. Fruit:
3-locular, fleshy or capsular, seeds small, rounded, with funicular
appendage at one end, or with a prominent appendage at both ends in
some species; seed coat dark brown, pitted, often sticky.
DISTRIBUTION   Tropical Asia.
POLLINATION   Some of the apostasioids are self-pollinating, but aside

from this we know nothing of the natural pollination of this group. Such knowledge, especially for *Neuwiedia*, might give us some hints as to the pollinator of the ancestral group from which the orchids have evolved. *Apostasia*, on the other hand, is not at all like any ancestral orchid in its flower structure.

CHROMOSOME NUMBER About 144 very small chromosomes (only one count published).

SPECIES About 16.

GENERA 2: *Apostasia, Neuwiedia.*

DISCUSSION This group was long classed as a tribe coordinate with the lady slippers in the subfamily Cypripedioideae. There are several divergent viewpoints on this group, some authors excluding them from the Orchidaceae largely because they do not look like other orchids or because they are not closely allied to the lady slippers. The apostasioids have slender style and equal and similar stigma lobes, features not found in other orchid groups (fig. 7.1). The apostasioid pollen occurs as monads, but it is not clear whether or not the pollen is sticky. Barthlott (1976b) suggests that their seed structure supports their exclusion from the Orchidaceae, and feels that the seed structure is more aberrant than that of *Selenipedium* or *Vanilla*. Most other features cited as arguments for their exclusion from the Orchidaceae do not hold up under close scrutiny. The style and filaments are distinctly united, the anthers are not basically different from those of other orchid groups, and the flowers are resupinate, with a definite lip in *Neuwiedia*. Some authors assign the aspostasioids a position near the Burmanniaceae or the Hypoxidaceae, but de Vogel (1969) indicates that there is no close relationship between the apostasioids and the Burmanniaceae, and that the resemblances to the Hypoxidaceae are quite superficial. Most authors who would exclude the apostasioids from the Orchidaceae still treat them as ancestral to the orchids. While the pollen structure of the apostasioids is distinctive and very different from that of the Cypripedioideae, Schill (1978) finds a resemblance in sculpturing between the apostasioids and some Spiranthoideae (See also Newton and Williams, 1978).

Unless better evidence for their exclusion is found, the best status for the apostasioids seems to be as a subfamily of the Orchidaceae. The symmetry of the anthers and the resupination of *Neuwiedia* both suggest a close relationship with the orchids. I would argue, however, that they are by no means to be considered as ancestral. Rather, they are a group in which the stigmatic fluid never functioned as an adhesive, and there was no evolution toward a rostellum, perhaps

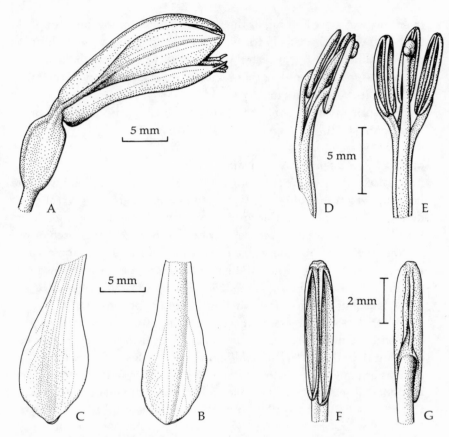

Figure 7.1 *Neuwiedia veratrifolia* (Apostasioideae). *(A)* Flower, side view. *(B)* Lip, flattened. *(C)* Petal, flattened. *(D)* Column, side view. *(E)* Column, ventral view. *(F)* Anther, ventral view. *(G)* Anther, dorsal view. Prepared from material preserved in liquid.

because the style projects beyond the anthers in most species.

RELATIONSHIPS In spite of having similar floral diagrams, the apostasioids are not closely allied to the lady slippers. The structure of the apostasioid flower and the numerous tiny chromosomes both argue against any close relationship. If there is a close relationship to any other living group, it is probably with the Spiranthoideae, and especially with the Tropidiinae. These two groups have similar habit, and *Apostasia* and some species of *Tropidia* also have similar storage roots. These resemblances need to be checked by careful study, especially of anatomy.

PHYLETIC TRENDS The genus *Apostasia* is specialized in the loss of the median anther and in the way in which the lateral anthers clasp the style.

REFERENCES   Larsen, 1969 (chromosome number); Newton and Williams, 1978 (pollen structure); Rao, 1969, 1974 (floral anatomy); Schill, 1978 (pollen structure); Siebe, 1903 (vegetative anatomy); de Vogel, 1969 (taxonomic revision).

Subfamily

CYPRIPEDIOIDEAE Lindley

DESCRIPTION   Habit: terrestrial, lithophytic, or epiphytic in humus; stems elongate or condensed, sometimes branched in *Selenipedium*, without pseudobulbs or storage roots. Leaves: spiral or distichous, convolute and plicate, or conduplicate and fleshy, nonarticulate, sometimes with pale blotches. Inflorescence: terminal, of one to several flowers; flowers spiral or distichous. Flowers: small to rather large, resupinate, with an articulation between ovary and perianth in the genera with conduplicate leaves; lateral sepals united; lip deeply saccate; lateral anthers fertile, median anther sterile and shieldlike; filaments largely united with style; pollen grains as monads, sticky and pastelike or united into pollinia (*Phragmipedium*); style thick, the stigma large and domelike. Fruit: 3-locular or 1-locular, capsular; seeds subspheric with a hard seed coat (*Selenipedium*) or minute with papery seed coat.

DISTRIBUTION   Tropical America, North America, Eurasia, and tropical Asia.

POLLINATION   All are trap flowers and, as far as known, none offers any reward. *Cypripedium* is pollinated by several different kinds of bee, and Vogel (1978) suggests that *C. debilis* may be pollinated by fungus flies. Both halictine bees and syrphid flies have been observed pollinating *Phragmipedium*. There are no data on the other genera.

CHROMOSOME NUMBERS   20, 24, 26, 28, 30, 32, 34, 36, 38, 40, 42. *Cypripedium* has 20 chromosomes, and the basic number of *Paphiopedilum* is 26, with aneuploidy and especially "centric fission," that is, breakage at the centromere to derive two pairs of telocentric chromosomes from one pair of metacentric chromosomes. The few counts for *Phragmipedium* are 20, 24, and 32. The chromosomes are relatively large in this subfamily.

SPECIES   About 115.

GENERA 4:   *Cypripedium, Paphiopedilum, Phragmipedium, Selenipedium.*

INTERGENERIC HYBRIDS   Crosses between *Phragmipedium and Paphiopedilum* are difficult to make, but two X *Phragmipaphium* crosses are

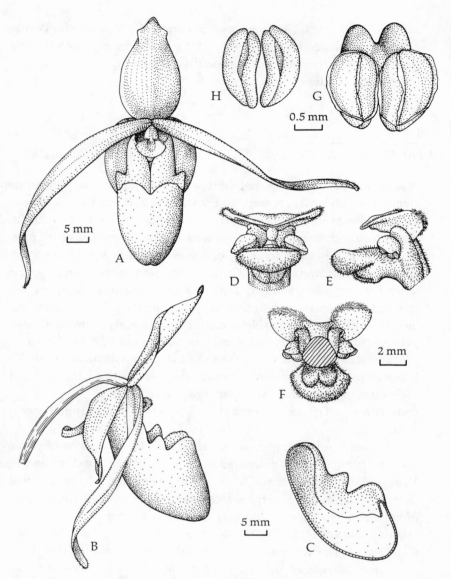

Figure 7.2 *Phragmipedium longifolium* (Cypripedioideae). *(A)* Flower, front view. *(B)* Flower, side view. *(C)* Longitudinal section of the lip. *(D)* Column, front view. *(E)* Column, side view. *(F)* Column, from base. *(G)* Anther. *(H)* Pollinia.

registered. As far as I know, neither one has been carefully studied to confirm its hybrid nature (see Wilson, 1961).

DISCUSSION   The stereotyped flower structure of this subfamily kept the Linnaean name *Cypripedium* in use for all subgroups for many years, but horticultural usage has adapted to more exact botanical knowledge, and all four genera are usually known by their correct names. The genera are distinctive, and Brieger actually uses four tribes and four subtribes within this subfamily (Schlechter, 1971). The four tribes seem excessive, but the subtribes are just silly. The tropical plants with conduplicate leaves are similar in tropical America (*Phragmipedium;* see fig. 7.2) and in tropical Asia *(Paphiopedilum)*, and it is not at all clear that they represent parallel evolution from a remote, plicate-leaved ancestor. The abscission layer between the ovary and the perianth is an unusual feature shared by the two genera, and they are anatomically very similar.

RELATIONSHIPS   Though a few authors have favored removal of this group from the Orchidaceae, it is closely related to the Neottieae-Limodorinae. The habit is very similar in *Cypripedium* and *Epipactis* or *Cephalanthera*. Both groups share unusually large chromosomes, and Barthlott finds the seed structure to be nearly identical (1976b). Pollen structure, however, is quite different in these groups. If we ever obtain a hybrid between two orchid subfamilies, I predict that it will be *Cypripedium* × *Epipactis*. Fruit and seed structure (and chemistry) suggests a link between *Selenipedium* and the Vanilleae; and that is also suggested by the abscission layer between the ovary and perianth in some lady slippers and most Vanillinae.

PHYLETIC TRENDS   The main trend within this subfamily is from the primitive habit with elongate stems and spiral leaves to a more condensed stem (within *Cypripedium*) and to distichous, conduplicate, and fleshy leaves (*Paphiopedilum* and *Phragmipedium*). Both genera with conduplicate leaves tend to be lithophytes and humus epiphytes.

REFERENCES   Dodson, 1966 (pollination); Garay, 1979 (review of *Phragmipedium*); Karasawa, 1979 (chromosomes of *Paphiopedilum*); Newton and Williams, 1978 (pollen structure); Pfitzer, 1903 (classification, needs revision); Rosso, 1966 (vegetative anatomy); Stoutamire, 1967 (pollination); Waters and Waters, 1973 (general).

The Spiranthoid and
Orchidoid Orchids

8

We will here consider two primarily terrestrial subfamilies (and some leftovers) with a single anther and soft pollinia. But once again these subfamilies are not at all closely related to each other, though some members of each group show about the same degree of evolutionary specialization. We tend to think of these orchids as just one step above the apostasioids and the lady slippers on the evolutionary ladder; however, some of the orchids in these two subfamilies are as specialized and complex in their floral features as any member of the Epidendroideae or the Vandoideae. These two subfamilies, which each include successful groups with hundreds of species, are by no means living fossils.

Subfamily

SPIRANTHOIDEAE Dressler, 1979b

DISCUSSION   The members of this subfamily are characterized especially by the dorsal, erect anther which is subequal to the rostellum and by the pollinia being attached to a viscidium at their apex. There are very few exceptions to the position of the viscidium, and these few are all clearly derived from the more usual pattern. In their basic column structure, then, these are sharply distinguished from the majority of the Orchidoideae, in which the anther projects beyond the stigma and the attachment of the pollinia to the viscidia is usually basal. These two groups are further differentiated in that the Spiranthoideae never have root-stem tuberoids, usually have definite mesoperigenous subsidiary cells accompanying the stomata (Williams, 1975), and have relatively small chromosomes. In the Orchidoideae, on the other hand, most genera have root-stem tuberoids (except in the Neottieae, which are unlikely to be confused with the Spiranthoideae), the stomata have no subsidiary cells, and the chromosomes are usually somewhat larger. In the Cranichideae, especially, the spiranthoid roots may be fleshy, but they are all similar on any one plant, never sharply differentiated into two kinds.

RELATIONSHIPS   The Spiranthoideae are without close allies among the other monandrous groups. However, the resemblances between the Tropidiinae and the apostasioids may represent a real phyletic relationship. The connection between the anther and the rostellum is consistently apical in the spiranthoids, and they could easily have evolved from *Apostasia*-like ancestors.

PHYLETIC TRENDS   The Tropidiinae seem to be close to the primitive growth habit, resembling the apostasioids. In their sectile pollinia, they show a definite alliance to the Goodyerinae. The Goodyerinae show a distinctive growth habit, with some specialization to epiphytism or to a storage rhizome (*Cheirostylis*). They are primarily herbs of wet forests, and the leaves often show interesting patterns of pink, silver, or pale green on a dark background (thus the "jewel orchids"). The ecological significance of these patterns is unknown. Their diversity in floral structure is more noteworthy. One may guess that a *Goodyera*-like flower with simple, saccate lip is close to the ancestral sort. Many genera have well-developed spurs, and the blade of the lip is often two-lobed. In some genera the blade of the claw may be deeply fringed (fig. 8.2). Some species of *Anoectochilus* show a curious retrorse appendage that extends into the spur from the column.

Tribe ERYTHRODEAE Dunsterville and Garay

DISCUSSION   The members of the Erythrodeae are clearly distinguished from the Cranichideae by the habit, in which the roots are scattered on an elongate rhizome rather than clustered, and by the sectile pollinia. In other respects the flower structure is similar in the two groups, and most authors have recognized a close relationship between them.

Subtribe TROPIDIINAE Pfitzer

DESCRIPTION   Habit: terrestrial, with slender, hard, rather woody, reed-like stems, up to 3 m in height, stem branched or not; roots sometimes with nodular storage tuberoids. Leaves: spiral or distichous, convolute, plicate, nonarticulate, scattered along the stem. Inflorescence: terminal or lateral, simple or branched, flowers spiral. Flowers: small or medium, resupinate; lip saccate or spurred basally or narrow and not saccate; column short or elongate, with the anther dorsal, erect, and subequal to the rostellum; two pollinia, sectile, attached to a terminal viscidium directly or by a single, slender caudicle; stigma entire.

DISTRIBUTION   Pantropical (see fig. 2.8).

POLLINATION   Not known. The structure of the long, pale flowers of *Corymborkis* suggests sphingid moth pollinations, as noted by Rasmussen (1977).

CHROMOSOME NUMBERS   40, 56, 60.

SPECIES   About 12.

GENERA   2: *Corymborkis, Tropidia.*

DISCUSSION   These two genera combine a rather primitive habit with column structure typical of the subfamily and such advanced features as lateral inflorescence and a well-developed caudicle in *Corymborkis,* and sectile pollinia in both genera. The caudicle has been interpreted as a stipe, but Rasmussen (1977) considers it to be derived from the anther. Still, such a single caudicle appears different from the caudicles of other Spiranthoideae, and it may be derived from the anther in a different way.

Preliminary observations of the stomatal pattern of *Corymborkis* suggest that the Tropidiinae may not be as closely allied to the Goodyerinae and Cranichideae as I have thought. If this is confirmed by further study, then this small group may be isolated, or possibly close to the apostasioids.

Butzin (1971) considers the name Corymbidinae Miquel to have priority over the Tropidiinae, but, as indicated by Rasmussen (1977), it is based on an illegitimate generic name and must be rejected.

REFERENCE   Rasmussen, 1977 (taxonomic revision of *Corymborkis*).

Subtribe GOODYERINAE Klotzsch

DESCRIPTION   Habit: terrestrial or infrequently saprophytic or epiphytic; rhizome creeping, occasionally thicker than leafy stem, leafy stem slender; roots somewhat fleshy, as ridges only in *Cheirostylis*. Leaves: spiral, scattered, or clustered, convolute, conduplicate, nonarticulate, often marked with pale or pink spots or lines. Inflorescence: terminal, unbranched, of few to many flowers, flowers spiral. Flowers: small or small-medium, usually white or pale green, usually resupinate; lip saccate at base, or forming a spur, often with emergent glands within the sack or spur, lip may be basally united with the column; blade often two-lobed, the claw or the blade sometimes fringed; anther dorsal, erect, subequal to rostellum; two or four pollinia, sectile, sometimes with two interlocular caudicles or with stipe(?); stigma entire or bilobed. (See fig. 8.1.)

DISTRIBUTION   Pantropical and northern, but mainly in tropical Asia.

POLLINATION   Bumblebee pollination is reported for *Goodyera*, and confirmed by Ackerman (1975) and Kallunki (1976), both of whom indicate

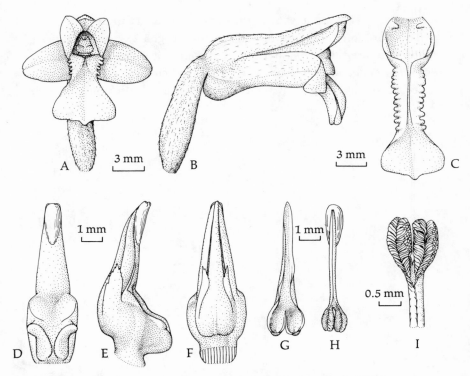

Figure 8.1 *Pristiglottis montana* (Spiranthoideae: Erythrodeae). *(A)* Flower, front view. *(B)* Flower, side view. *(C)* Lip. *(D)* Column, ventral view. *(E)* Column, side view. *(F)* Column, dorsal view. *(G)* Anther. *(H)* Pollinarium. *(I)* Pollinia.

that fruit-set is usually very good. Butterfly pollination is reported for *Ludisia*. Many of the Asiatic genera appear to be adapted for pollination by Lepidoptera.

CHROMOSOME NUMBERS   20, 22, 26, 28, 30, 32, 40, 42, 44, 56.

SPECIES   About 425.

GENERA   36 in two alliances: (1) With a single stigmatic area: *Aspidogyne, Cystorchis, Dicerostylis, Dossinia, Erythrodes, Eurycentrum, Evrardia, Gonatostylis, Goodyera, Herpysma, Hylophila, Kreodanthus, Kuhlhasseltia, Lepidogyne, Ligeophila, Ludisia, Macodes, Moerenhoutia, Orchipedum, Papuaea, Platylepis, Platythelys, Pristiglottis, Rhamphorhynchus, Stephanothelys, Vieillardorchis.* (2) With two distinct stigmatic areas: *Anoectochilus, Chamaegastrodia, Cheirostylis, Eucosia, Gymnochilus, Hetaeria, Myrmechis, Tubilabium, Vrydagzynea, Zeuxine.*

INTERGENERIC HYBRIDS   Artificial crosses have been registered between *Ludisia* and *Anoectochilus, Dossinia* and *Macodes.*

DISCUSSION This group shows considerable diversity in tropical Asia, with elaboration of lip and column structure (fig. 8.2). Some authors continue to use the name Physurinae, though it is clearly based on a *nomen nudem* and should be rejected. I have listed the genera in two series, following Brieger (in Schlechter, 1974), but with a modification

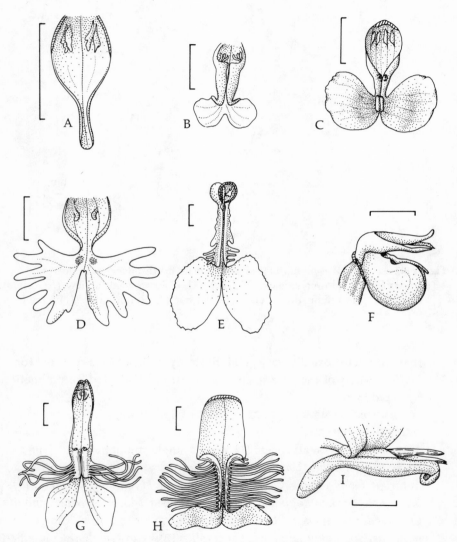

Figure 8.2 Representative lips of Asiatic Goodyerinae. *(A) Hetaeria obliqua. (B) Zeuxine nervosa. (C) Hetaeria rotundiloba. (D) Cheirostylis flabellata. (E) Anoectochilus brevistylis. (F) Anoectochilus siamensis. (G) Anoectochilus albolineata. (H) Erythrodes herpysmoides,* side view with column. *(I) Dicerostylis lanceolata,* side view with column. Scale 3 mm. (After Seidenfaden, 1978a.)

indicated by Seidenfaden (1978a). As Seidenfaden indicates, the genera of this group need revision.

REFERENCES   Ackerman, 1975 (floral biology of *Goodyera*); Garay, 1977 (delimitation of American genera); Kallunki, 1976 (North American *Goodyera*).

Tribe CRANICHIDEAE Endlicher

DISCUSSION   The Cranichideae show column structure similar to that of the Erythrodeae, and a stereotyped habit. They are all rosette plants, and the leafy stem and the rhizome are normally very condensed, so that the roots are clustered. There are a few exceptions, with more elongate rhizomes, but none with a habit like that of the Goodyerinae; and the pollinia of this group are never sectile. The name Spirantheae Endlicher is used by Brieger (Schlechter, 1971-1974), but Cranichideae had been used already in this inclusive sense by Dunsterville and Garay in 1965 (Dunsterville and Garay, 1959-76, vol. 3).

PHYLETIC TRENDS   This tribe is vegetatively uniform, though some Spiranthinae and a few *Ponthieva* show tendencies to epiphytism. Within the Spiranthinae there is some variation in the structure of the viscidia, probably reflecting in part adaptations to different types of pollinators, but it is not clear what these relationships may be. As in the Goodyerinae, there are genera with the base of the lip only slightly saccate, and others with well-developed nectaries. In some genera the nectary is united with the ovary, and not very obvious except in section, as in *Sarcoglottis*. In other cases, as in *Eltroplectis* and *Pelexia*, there is an obvious external chin or spur. The Cranichidinae show greater variation in floral structure, especially in the *Cranichis* alliance. The genera of the *Altensteinia* alliance, some of which occur at high elevations, are relatively uniform, but some species of *Myrosmodes* show a long "neck," which presumably represents a nectary within the flower. Two interesting trends occur in the *Cranichis* alliance. In *Ponthieva* especially, the lip, which is partly united with the column, is small and inconspicuous, while the two petals together form a sort of pseudolip, thus giving the aspect of resupination to these nonresupinate flowers. In *Pseudocentrum* and *Solenocentrum* we find well-developed spurs, a condition approached by a few species of *Cranichis*. *Cryptostylis* has adapted to pseudocopulation and is thus quite distinctive in its floral features.

Subtribe SPIRANTHINAE Lindley

DESCRIPTION   Habit: terrestrial or occasionally epiphytic; the leafy stem and rhizome short, roots clustered, usually fleshy. Leaves: spiral, clustered, convolute, conduplicate, commonly petiolate, nonarticulate. Inflorescence: terminal, with several to many spiral flowers. Flowers: small to medium, resupinate; lip basally saccate or not, base of the blade commonly with two retrorse lobules or appendages; flowers commonly with a deep nectary united with the ovary, this sometimes with a prominent chin or spur; column usually erect; anther dorsal, normally erect, and subequal to the rostellum; four or two pollinia, soft and mealy, the viscidium terminal or rarely attached to middle of pollinia; caudicles weakly developed in a few cases; stigma entire or two-lobed. (See fig. 8.3.)

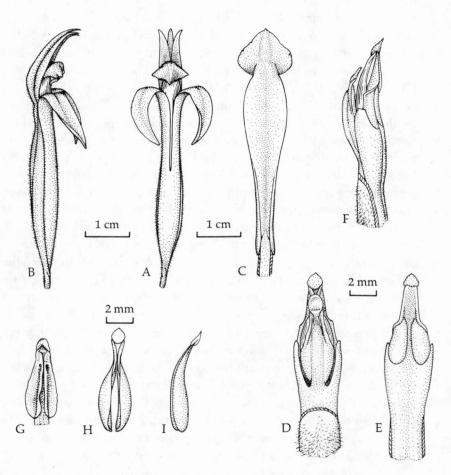

Figure 8.3 *Sarcoglottis acaulis* (Spiranthoideae: Cranichideae). (*A*) Flower, front view. (*B*) Flower, side view. (*C*) Lip, flattened. (*D*) Column, dorsal view. (*E*) column, ventral view. (*F*) Column, side view. (*G*) Anther. (*H*) Pollinarium, dorsal view. (*I*) Pollinarium, side view.

DISTRIBUTION   Primarily tropical American, but with a few representatives in all habitable continents except tropical and southern Africa.

POLLINATION   *Spiranthes* is reportedly pollinated by bumblebees, halictid bees, and megachilid bees. I have observed bumblebees pollinating *Pelexia ekmanii* in southern Brazil. *Stenorrhynchus* shows the features typical of hummingbird pollination. Most Spiranthinae are probably bee pollinated, and most of them seem to be well pollinated under natural conditions.

CHROMOSOME NUMBERS   24, 26, 30, 44.

SPECIES   About 275.

GENERA   28: *Beadlea, Beloglottis, Brachystele, Buchtienia, Coccineorchis, Cybebus, Cyclopogon, Deiregyne, Discyphus, Eltroplectis, Eurystyles, Funkiella, Galeottiella, Gamosepalum, Hapalorchis, Lankesterella, Lyroglossa, Mesadenella, Mesadenus, Pelexia, Pseudogoodyera, Pteroglossa, Sarcoglottis, Sauroglossum, Schiedeella, Spiranthes, Stenorrhynchus, Synassa.*

DISCUSSION   Schlechter (1920) divided this group into 24 genera, basing his classification primarily on the nature of the rostellum and the viscidium. A few genera have been added since then. Some authors go to the other extreme and include nearly all the genera in *Spiranthes,* which seems unrealistic, but Schlechter's system does seem rather finely split. A careful study of this group is needed, and the genera listed above may not all survive such a study. *Eurystyles* and *Lankesterella* are placed in different groups by Schlechter because of the nature of the rostellum. These two groups of small epiphytes are very similar in all other features, which casts some doubt on Schlechter's four series.

Though spotting and striping of the leaves are less frequent than in the Goodyerinae, some of the Spiranthinae show interesting variegation. In many cases they are polymorphic, with two or more patterns mixed within the population.

REFERENCES   Balogh, 1979 (pollinia); Schlechter, 1920 (generic classification).

Subtribe PACHYPLECTRONINAE Schlechter

DESCRIPTION   Habit: terrestrial; the leafy stem and rhizome short; roots clustered or somewhat scattered, fleshy. Leaves: spiral, clustered, convolute, conduplicate, petiolate, nonarticulate. Inflorescence: terminal, with several to many spiral flowers. Flowers: small, resupinate; lip with a prominent basal spur; column with prominent staminodia which enfold the sides of the anther, anther erect; two pollinia, soft and mealy, with a distinct viscidium; stigma entire.

DISTRIBUTION   New Caledonia.
POLLINATION   Not known.
CHROMOSOME NUMBERS   Not known.
SPECIES   2.
GENUS   *Pachyplectron.*
DISCUSSION   Kränzlin (1928) considered this genus referable to *Ery-throdes*, but neither the habit nor the column structure agrees with *Erythrodes*, and the pollinia are not sectile. The stomata of this genus have mesoperigenous subsidiary cells, and the plant and flowers resemble *Manniella* in several features, even though the spur of *Pachyplectron* is quite unlike the nectary of *Manniella*. Thus, I assign the Pachyplectroninae to the Cranichideae and place it near the Manniellinae.
REFERENCE   Hallé, 1977 (illustrations and descriptions).

Subtribe MANNIELLINAE Schltr.

DESCRIPTION   Habit: terrestrial, the leafy stem and rhizome relatively short, roots clustered, not very fleshy. Leaves: spiral, convolute, conduplicate, nonarticulate, petiolate. Inflorescence: terminal, of many spiral flowers. Flowers: small, resupinate, each with a prominent cuniculus; lip simple, with two retrorse basal lobules; column bent sharply up and down again (rather like a door latch), with two prominent staminodia which clasp the anther laterally; anther dorsal, erect; two pollinia, soft and mealy, viscidium terminal; stigma entire.
DISTRIBUTION   West tropical Africa.
POLLINATION   Not known.
CHROMOSOME NUMBERS   Not known.
SPECIES   1.
GENUS   *Manniella.*
DISCUSSION   This genus is clearly allied to the Spiranthinae and was included in that group by Mansfeld (1937a). However, the oddly bent column and the prominent staminodia are distinctive. Species of other areas have been assigned to *Manniella* because of united sepals and petals, but this is one-character taxonomy carried to a ridiculous extreme, as these species do not show the other distinctive features of *Manniella*.

Indeed, on studying Hallé's excellent drawings of *Manniella*, it is not clear that there is really any union between the sepals and petals of this genus. The base of the perianth is strongly oblique, as in *Pachyplectron* and many Spiranthinae. At the same time, a cuniculus-like nectary extends well below the base of the lip, as in *Sarcoglottis*. While the whole "neck" between the ovary and the free perianth parts is, in a strict morphological sense, a perianth tube with the column united to one side of it, the bases of the sepals and petals, as we usually interpret them, seem to be at or very near the upper end of this neck. The sepals and petals are united only in the sense that those of *Brassavola* are united. REFERENCE   Hallé, 1965 (illustration).

Subtribe CRANICHIDINAE Lindley

DESCRIPTION   Habit: terrestrial or lithophytic, rarely epiphytic; the leafy stem and rhizome short, roots clustered, rather fleshy. Leaves: spiral, clustered, convolute, conduplicate or subplicate, nonarticulate, commonly petiolate. Inflorescence: terminal or occasionally lateral, with several to many spiral flowers. Flowers: small or medium, nonresupinate; lip often saccate, sometimes united to column, or spurred; column straight or bent; anther dorsal, erect, subequal to rostellum; four pollinia, soft and mealy, or sometimes rather hard, sometimes with two or four terminal caudicles; viscidium terminal; stigma entire.

DISTRIBUTION   Tropical America, with a few representatives in North America, and one in New Caledonia.

POLLINATION   Not known. *Porphyrostachys pilifera* is presumably pollinated by hummingbirds.

CHROMOSOME NUMBER   Not known.

SPECIES   About 200.

GENERA   15 in two alliances: (1) Column blunt; pollinia soft, without caudicles; lip simple, free from column: *Aa, Altensteinia, Gomphichis, Myrosmodes, Porphyrostachys, Prescottia, Stenoptera.* (2) Column sharply pointed; pollinia firm, with caudicles; lip often united with column: *Baskervilla, Coilostylis, Cranichis, Fuertesiella, Ponthieva, Pseudocentrum, Pterichis, Solenocentrum.*

DISCUSSION   Brieger divides the group along different lines from the two alliances given here, but his system requires a distinction between columnar tissue and staminodial tissue that I am unable to see.

REFERENCE   Renz, 1948 (taxonomy).

Subtribe CRYPTOSTYLIDINAE Schlechter

DESCRIPTION   Habit: terrestrial or saprophytic; leafy stem and rhizome short, roots clustered, fleshy. Leaves: few, spiral, clustered, convolute, conduplicate, nonarticulate, somewhat petiolate. Inflorescence: terminal, simple, of few to many spiral flowers. Flowers: small to medium, nonresupinate, lip simple; column very short, anther dorsal, erect, subequal to rostellum; four pollinia, soft and mealy, with a terminal viscidium, without caudicles; stigma entire.

DISTRIBUTION   Australasia and tropical Asia.

POLLINATION   The genus *Cryptostylis* is one of the better known cases of pseudocopulation, and is interesting in that the *Cryptostylis* are all pollinated by the same species of ichneumon wasp in all recorded cases. As with *Ophrys*, some species usually lead the pollinator to contact the viscidium with its head and others cause the insect to place its abdomen against the column. The main barrier between these species, however, appears to be in the form of interspecific incompatibility.

CHROMOSOME NUMBER   42.

SPECIES   About 15.

GENUS   *Cryptostylis.*

DISCUSSION   Some authors assign this genus to the Cranichidinae, where it could be placed as a third alliance.

REFERENCE   Cady, 1967 (taxonomy of Australian species).

Subfamily

ORCHIDOIDEAE

DISCUSSION   This group is characterized by the usually erect anther that projects beyond the stigma, by a lack of subsidiary cells, by the habit of growth, and by the presence of the unique root-stem tuberoids in most members of the tribes Diurideae, Orchideae, and Diseae.

Most classifications have treated the Orchideae and Diseae as having no close allies, but several features point to a phyletic relationship between the Diurideae and the Orchideae and Diseae; together with the Neottieae, they make a meaningful phyletic group. The rostellum of the Orchideae is homologous with that of the other monandrous orchids, though the viscidium of this tribe is basically two-parted. Several of the features that have been held to prove the isolated position of the Orchideae and Diseae—the root-stem tuberoids, sectile pollinia, and basitony—are all shown by different members of the Diurideae. The interlocular caudicles and the double viscidium are not shown by the Diurideae (nor by all Orchideae), but these are evidence of the greater specialization of the Orchideae rather than evidence against a phyletic relationship. The occurrence of a putative hybrid between *Epipactis* and *Gymnadenia* also indicates a closer relationship between Neottieae and Orchideae than is shown by most systems. I interpret the Neottieae as a relic northern group. The Orchideae is basically an African group which has radiated outward to Eurasia and to a lesser degree to other continents. I interpret the Diurideae as a basically southern group that was split from the Orchideae-Diseae in the early Tertiary by the separation of Africa and

South America. The Chloraeinae represent those that remained in South America, while others reached Australia via Antarctica (then much more hospitable) and there radiated and diversified much more than in South America. The Antarctic corridor was probably habitable for some plants well into the mid Tertiary. The much greater diversity of the Diurideae in Australia suggests that the group might have spread from Australia to South America, and this might have been the case, though Australia is thought to have been separated from Africa well before the separation of South America from Africa.

RELATIONSHIPS   As one of the major lines of orchid evolution, this is a distinctive group, but the Neottieae seem to show a close alliance to the lady slippers. This subfamily is also, but less obviously, related to the Epidendroideae and the Vandoideae.

Tribe NEOTTIEAE Lindley

DISCUSSION   This tribe includes the most primitive genera of the Orchidoid line of evolution. The habit of the Limodorinae is quite comparable to that of *Cypripedium* and some other very primitive orchids.

PHYLETIC TRENDS   Quite a few members of the Neottieae are saprophytes. The Limodorinae, especially, are relatively primitive in both habit and flower structure. The Listerinae are more advanced in both the condensed vegetative habit and in the nature of the rostellum.

Subtribe LIMODORINAE Bentham

DESCRIPTION   Habit: terrestrial or saprophytic, with elongate stems and more or less clustered roots, roots slender but somewhat fleshy. Leaves: spiral, scattered on stem, convolute, plicate, nonarticulate. Inflorescence: terminal, of several of many spiral flowers, simple. Flowers: small to medium; mid-lobe of lip often hinged to a more or less saccate base or lip with a spur; column short or long, the anther dorsal, erect, extending beyond the stigma or rostellum; two pollinia, soft and mealy, without a distinct viscidium; stigma entire. (See fig. 8.4.)

DISTRIBUTION   Tropical Asia, northern hemisphere, and tropical Africa.

POLLINATION   We have good data for some genera of the Neottieae, as they occur in Europe. *Cephalanthera* and *Limodorum* are reportedly pollinated by bees, while most species of *Epipactis* are wasp flowers. However, *E. gigantea* is said to be pollinated by syrphid flies, and *E. atrorubens* and *E. palustris* are usually pollinated by bees (see Wiefelspütz in Senghas and Sundermann, 1970). In all cases, these flowers seem to offer nectar in open nectaries. Ivri and Dafni (1977) find that *Epipactis con-*

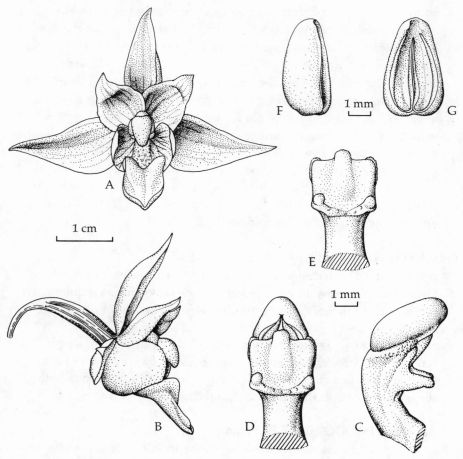

Figure 8.4 *Epipactis gigantea* (Orchidoideae: Neottieae). *(A)* Flower, front view. *(B)* Flower, side view. *(C)* Column, side view. *(D)* column, ventral view, with anther in place. *(E)* Column ventral view, with anther removed. *(F)* Anther, side view. *(G)* Anther, ventral view. Drawn from material preserved in liquid.

similis is pollinated by syrphid flies, and they suggest that warts on the lip of this species mimic aphids and help attract female syrphids that normally lay their eggs among aphids.

CHROMOSOME NUMBERS   32, 34, 36, 40, 48, 64.

SPECIES   About 60.

GENERA 5:   *Aphyllorchis, Cephalanthera, Epipactis, Limodorum, Thaia.*

INTERGENERIC HYBRIDS   Although *Cephalanthera* and *Epipactis* are different in column structure, several natural intergeneric hybrids are recorded.

DISCUSSION   This rather primitive group shows a strong tendency to saprophytism and self-pollination. *Aphyllorchis* species should all be considered critically, as saprophytes from quite different groups may be hiding here. The Australian *Aphyllorchis anomala*, for example, shows not only a lip structure atypical for the genus but what appears to be a totally different anther structure. It may well belong to a different group.

Chen (1965) has described a Chinese orchid as *Tangtsinia*, which he interprets as a primitive genus. I would judge that it is rather a peloric mutant of *Cephalanthera*.

REFERENCES   Chen, 1965 *(Tangtsinia)*; Senghas and Sundermann, 1970 *(Epipactis)*.

Subtribe LISTERINAE Lindley

DESCRIPTION   Habit: terrestrial or saprophytic; small herbs with slender stems; roots fleshy but slender. Leaves: usually two, subopposite, convolute, conduplicate, nonarticulate. Inflorescence: terminal, simple, of several to many spiral flowers. Flowers: small, usually greenish; lip with a shallow, basal nectary; column short, anther suberect, subequal to the rostellum; two pollinia, soft and mealy; stigma entire, rostellum sensitive, extruding a drop of adhesive when touched.

DISTRIBUTION   Northern.

POLLINATION   The genus *Listera* is primarily pollinated by nectar-seeking wasps and primitive flies.

CHROMOSOME NUMBERS   34, 36, 38, 40, 42, 56.

SPECIES   About 38.

GENERA 2:   *Listera, Neottia.*

REFERENCES   Ackerman and Mesler, 1979 (pollination of *Listera*); Ackerman and Williams, in press *a* (pollen structure); Ramsey, 1950 (rostellum).

Tribe DIURIDEAE Endlicher

DISCUSSION   This primarily Australian group has radiated and diversified in habit and flower structure, but it remains a natural group, characterized especially by root-stem tuberoids (in most Australian genera). Lavarack (1970, 1976) has written a useful study of this group, and the subtribes used here follow his treatment for the most part. I have added the South American Chloraeinae and reordered the subtribes to place the Caladeniinae after the Chloraeinae and before the more specialized groups. Lavarack notes that the tribe is best developed in the nontropical parts of Australia, and that the plants are characteristic of open, relatively dry habitats.

The Australian genera seem to form two major complexes: one with short column, prominent staminodia, and usually narrow leaves, the Diuridinae and Prasophyllinae, and the other subtribes, usually with longer column, without prominent staminodia, and with wider leaves. Barthlott (pers. comm.) reports that these two complexes are distinctive in their seed structure. Brieger (in Schlechter, 1974, 1975) places the Chloraeinae with the Spirantheae (= Cranichideae) because of their root structure (clustered, fleshy roots, rather than root-stem tuberoids). However, column structure, pollen structure, and general flower structure all show a close alliance between the Chloraeinae and the Australian genera.

PHYLETIC TRENDS   The Diurideae are rather diverse, with the Chloraeinae being relatively primitive and similar to the Neottieae. Many Australian Diurideae have adapted to pseudocopulation and highly seasonal environments, so that there is a good deal of vegetative diversity and floral specialization. Viscidia are present in many Diurideae, and a stipe is found in *Prasophyllum*. Several genera have evolved motile lips which trap the pollinators or throw them against the column.

REFERENCES   Ackerman and Williams, in press b (pollen structure); Lavarack, 1970, 1976 (classification of Australian members).

Subtribe CHLORAEINAE Reichenbach fil.

DESCRIPTION   Habit: terrestrial or rarely epiphytic, roots fleshy, or with root-stem tuberoids (*Codonorchis*), stems slender. Leaves: spiral, clustered or scattered on stem, whorled in middle of stem in *Codonorchis*, convolute, conduplicate, nonarticulate. Inflorescence: terminal, simple, of one to many spiral flowers. Flowers: medium, often with osmophores on lateral sepals; lip usually adorned with warts or calluses; may have two slender nectaries in the base of the flower (*Chloraea*); column slender, usually arched; anther dorsal, erect, projecting beyond the rostellum; two pollinia, of tetrads or monads, soft and mealy, without viscidium; stigma entire. (See fig. 8.5.)

DISTRIBUTION   Southern South America north to Peru, and New Caledonia.

POLLINATION   Gumprecht (1975) reports the pollination of *Chloraea* by *Colletes*, a rather primitive bee.

CHROMOSOME NUMBER   Not known.

SPECIES   About 100.

GENERA   6: *Bipinnula, Chloraea, Codonorchis, Gavilea, Geoblasta, Megastylis*.

DISCUSSION   No South American material was available to Lavarack, and so he could not judge the relationship of this subtribe to the Australian

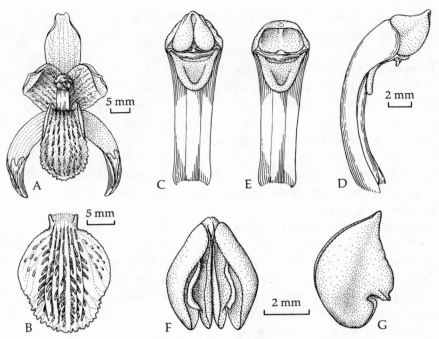

Figure 8.5 *Chloraea lamellata* (Orchidoideae: Diurideae). (*A*) Flower, front view. (*B*) Lip, flattened. (*C*) Column, ventral view. (*D*) Column, side view. (*E*) Column, with anther removed. (*F*) Anther, ventral view, (*G*) Anther, side view. Drawn from material preserved in liquid.

genera. The root-stem tuberoids, which characterize nearly all the Australian genera, are lacking in most Chloraeinae but are present in *Codonorchis*. Further, the flower structure of the South American genera is very similar to that of the Australian groups. The column of *Chloraea* closely resembles that of *Burnettia, Lyperanthus,* and *Rimacola,* for example. A recent study of pollen structure by Ackerman and Williams (in press b) indicates a strong relationship between the Chloraeinae and the Australian members of this tribe. *Codonorchis* is distinctive in both habit and flower structure. Pfitzer (1887) assigns it to the Caladeniinae, but I would judge that *Chloraea* itself is closer to the Caladeniinae than is *Codonorchis*. Brieger (in Schlechter, 1975) suggests that *Codonorchis* should form a separate subtribe, but I see little need for such a separation. It seems to be close to *Gavilea* in lip structure. The New Caledonian *Megastylis* has been treated as a distinct subtribe by Schlechter (1926), but seems very close to the South American genera in all features. *Gavilea odoratissima* is usually epiphytic, the only epiphytic orchid reported for Chile (Riveros and Ramírez, 1978).

REFERENCES   Correa, 1968 (*Geoblasta*), 1969 (taxonomic revision of *Chloraea*); Gumprecht, 1975 (general and pollination).

Subtribe CALADENIINAE Pfitzer

DESCRIPTION   Habit: terrestrial, with root-stem tuberoids. Leaves: few, basal, convolute, conduplicate, nonarticulate. Inflorescence: terminal, simple, of one to several spiral flowers. Flowers: resupinate, small to medium; lip usually adorned with warts, hairs, or calluses, sometimes hinged or sensitive; column arched, often broadly winged above; anther conic, dorsal, erect, projecting beyond the rostellum; four or eight pollinia, of monads or tetrads, soft and mealy, stigma entire, with or without a distinct viscidium.

DISTRIBUTION   Australasia and into tropical Asia.

POLLINATION   The genera *Caladenia, Chiloglottis, Drakaea,* and *Spiculaea* are all known to be pollinated through pseudocopulation by male thynnid wasps, as described in chapter 4. We may guess that *Caleana* and *Paracaleana* show a similar relationship. We have no further data on pollination in this subtribe.

CHROMOSOME NUMBER   Not known.

SPECIES   About 100.

GENERA   16 in two alliances: (1) Lip not motile: *Adenochilus, Aporostylis, Burnettia, Caladenia, Chiloglottis, Elythranthera, Eriochilus, Glossodia, Leporella, Lyperanthus, Rimacola.* (2) Lip hinged and motile: *Arthrochilus, Caleana, Drakaea, Paracaleana, Spiculaea.*

DISCUSSION   This subtribe is characterized by flat, more or less basal leaves, by an arched column that may be broadly winged above but does not have staminodia-like wings as do most Diuridinae. The lip is generally ornamented with warts or calluses. We may guess that something like *Burnettia, Lyperanthus,* or *Rimacola,* ornate as they are, are probably close to the primitive form for this subtribe. The warty excrescences on the lip predisposed this group for the evolution of pseudocopulation by thynnid wasps. This has lead to such bizarre flowers as *Spiculaea* and *Drakaea.*

Spiculaea has carried the adaptation to dry habitats to an extreme degree. In some species, at least, if the plant is cut or pulled up at flowering time, it is able to ripen seeds with the water and foodstuffs available in the shoot (Northen, 1971). Presumably, the tuberoid also retains enough reserves to produce a new shoot the next year. Brieger (in Schlechter, 1975) places *Rimacola* with the Chloraeinae, as it lacks root-stem tuberoids. In all other features, though, it is more similar to the other Australian genera.

REFERENCES   Blaxell, 1972 (taxonomy of *Caleana* alliance); Cady, 1962 (taxonomy of *Spiculaea*); Northen, 1971 (survival of *Spiculaea*); Rotherham, 1968 (pollination of *Spiculaea*).

Subtribe PTEROSTYLIDINAE Pfitzer

DESCRIPTION   Habit: terrestrial with root-stem tuberoids. Leaves: several, spiral, clustered or scattered, broad, convolute, conduplicate, nonarticulate. Inflorescence: terminal, simple, of one to several spiral flowers. Flowers: small to medium; resupinate; the dorsal sepal commonly hood-like, and the laterals often forming slender tails; lip narrow, hinged, with a basal appendage, sensitive; column arched, slender below, winged above, the wings with projections both distally and basally; anther dorsal, more or less erect; four pollinia, soft and mealy, of monads, without viscidium; stigma entire.

DISTRIBUTION   Australasia.

POLLINATION   The green hoods form fly traps, and the sensitive lips throw small flies against the columns.

CHROMOSOME NUMBER   Not known.

SPECIES   About 71.

GENUS   *Pterostylis*.

DISCUSSION   The "greenhoods" make up the largest of the primarily Australian genera. *Pterostylis* is very distinctive; the sepals generally form a hood, from one side of which the narrow lip may project before it is tripped by a visiting gnat. The pollinia are quite soft and mealy, possibly an adaptation to pollination by small flies, though Ackerman and Williams (in press *b*) interpret this as a primitive feature.

REFERENCES   Cady, 1969 (checklist); Northen, 1972 (pollination mechanism).

Subtribe ACIANTHINAE Schlechter

DESCRIPTION   Habit: terrestrial, with root-stem tuberoids. Leaves: basal, solitary, broad and cordate or lobed, convolute, conduplicate or plicate, nonarticulate. Inflorescence: terminal, simple, of one to several spiral flowers. Flowers: small or medium, resupinate; column slender and arched or short and relatively thick, without wings; anther terminal; pollinia of monads or tetrads, four, in two pairs, with a viscidium; stigma entire.

DISTRIBUTION   Australasia and tropical Asia.

POLLINATION   Both *Acianthus* and *Corybas* are reported to be pollinated by fungus flies. *Acianthus* is a good example of a flower which advertises nectar with a (to us) unpleasant odor (D. L. Jones, 1974). Vogel (1978) suggests that the flowers of *Corybas* are fungus mimics.

CHROMOSOME NUMBER   Not known.

SPECIES   About 100.

GENERA   4: *Acianthus, Corybas, Stigmatodactylus, Townsonia*.

DISCUSSION   The odd, *Asarum*-like flower of *Corybas* is made up primarily of the lip and the dorsal sepal, the other segments being much smaller or rudimentary. The genus seems to have found a niche in wet forests to a greater degree than any other member of the Diurideae, and it is well distributed in Asia. At the same time, *C. macrantha* recently went on record as being the southernmost orchid, when it was found on subantarctic Macquarie Island (Brown et al., 1978). Vegetatively, *Acianthus* and *Townsonia* are very much like *Corybas*, with a single, more or less heart-shaped leaf. However, the flower structure and the pollen structure are so different that one is tempted to separate *Corybas* from this subtribe (as was done by Schlechter, 1926). In some species of *Acianthus* the anther is conic and projects beyond the rostellum, as in most Chloraeinae and Caladeniinae. In most species, however, the anther is more or less terminal, short, and firmly united with the column, and opens away from the clinandrium, as in the Orchideae or Sunipiinae. *Stigmatodactylus* is saprophytic and ranges into Asia. Published illustrations of *Stigmatodactylus* suggest superposed pollinia, but I have seen no good material, and I am unsure of the relationship between this genus and *Acianthus*.

REFERENCE   D. L. Jones, 1974 (pollination of *Acianthus*).

Subtribe DIURIDINAE Lindley

DESCRIPTION   Habit: terrestrial, with root-stem tuberoids. Leaves: few to several, largely basal, convolute, conduplicate, usually narrow or rather fleshy, nonarticulate. Inflorescence: terminal, simple, of several to many spiral flowers. Flowers: small to medium; column short, generally bearing prominent, armlike wings or staminodia parallel with the column or projecting beyond it; anther conic, dorsal, erect or somewhat bent forward, projecting somewhat beyond the rostellum; four pollinia, of monads or tetrads, soft and mealy, generally with a terminal viscidium; stigma entire.

DISTRIBUTION   Australasia and into tropical Asia.

POLLINATION   The genus *Diuris* is reportedly pollinated by small bees, and *Calochilus* is pollinated through pseudocopulation by male wasps. We have no data for the other genera.

CHROMOSOME NUMBER   26.

SPECIES   About 90.

GENERA   5: *Calochilus, Diuris, Epiblema, Orthoceras, Thelymitra.*

DISCUSSION   This group is characterized by the short and generally much-ornamented column. The viscidium is quite terminal on the pollinia in this group. *Diuris* is especially interesting in that one can scarcely speak of a column. There is virtually no union between the style and either the very short filament or the staminodia. The column of *Diuris* seems primitive, but it may only be very short, and the perianth is quite unusual. We have, as yet, no idea what has selected for the not-very-orchidlike flowers of *Thelymitra.*

REFERENCE   Jones and Gray, 1974 (pollination of *Calochilus*).

Subtribe PRASOPHYLLINAE Schlechter

DESCRIPTION   Habit: terrestrial, with root-stem tuberoids. Leaves: few, basal or nearly so, narrow or cylindrical, nonarticulate. Inflorescence: terminal, simple, of few to many spiral flowers. Flowers: small to medium, resupinate or not; column short, with two prominent staminode-like wings; anther dorsal, either shorter or longer than the rostellum; two or four pollinia, sectile, with a basal viscidium or with a terminal viscidium and a stipe; stigma entire.

DISTRIBUTION   Australasia and into tropical Asia.

POLLINATION   The species *Microtis parviflora* is reported to be pollinated by ants, but self-pollinates if it is not pollinated (Jones, 1975).

CHROMOSOME NUMBER   44.

SPECIES   About 90.

GENERA   3: *Genoplesium, Microtis, Prasophyllum.*

DISCUSSION   This group, with rather onionlike leaves, is characterized by sectile pollinia. Having sectile pollinia and a basitonic viscidium, *Microtis* is more Orchideae-like than any other of the Australian genera. This, however, is probably parallelism (from similar ancestors) rather than an indication of especially close relationship. *Prasophyllum* is interesting for the independent evolution of a prominent stipe. At first glance, the anther structure of these two genera seems too unlike to be placed in the same subtribe, but the variation in column structure of *Prasophyllum* seems to bridge the gap.

REFERENCE   Jones, 1975 (pollination of *Microtis*).

Tribe ORCHIDEAE

DISCUSSION   This orchid group, which includes the type genus for the family, is well represented in Europe, and for this reason we know a great deal more about this than most other orchid tribes. Nevertheless, it is basically an African group. The Huttonaeinae are exclusively African, and the Habenariinae are largely African, though a few genera are widespread. The Orchidinae, too, are represented in Africa, and the African genera are, in general, more distinctive than those of Eurasia. One would guess that the Orchidinae must have invaded Eurasia fairly early in the Tertiary to have diversified so much.

The Orchideae, with the Diseae, form one of the most distinctive groups in the family, and many authors consider this group to be very isolated. The Orchideae are characterized by root-stem tuberoids, by sectile pollinia with basal, interlocular caudicles, by the complete union of the base of the anther with the column, and by the basically two-parted viscidium. The anther is erect in the Orchideae, while the Diseae have the anther bent back or even upside down. (See fig. 8.6.)

I have chosen to separate the Diseae from the Orchideae, through uniting both under the Orchideae would emphasize the unity of the group. Neither classification would be in any way unnatural, according to presently available data. The division of the Orchideae into sub-tribes, however, is not altogether convincing, and I am sure that this group is in need of study on a worldwide basis.

PHYLETIC TRENDS. Unlike the Diseae, the ancestral Orchideae developed a spur at the base of the lip rather early in their evolution. *Ophrys* has lost its spur, if it ever had one, and the relative separation of anther and column suggests that *Ophrys* may have preserved some rather primitive features. Another genus that is suggestive of a generalized sort of Orchidinae is *Brachycorythis*. One of the prominent trends has been the great development of the spur, along with the development of long-stalked pollinia and long-stalked stigmas.

Subtribe ORCHIDINAE

DESCRIPTION Habit: terrestrial, rarely saprophytic, usually with root-stem tuberoids; stem slender. Leaves: spiral, scattered or basal and clustered, convolute, conduplicate, nonarticulate. Inflorescence: terminal, simple, of few to many spiral flowers. Flowers: small to medium-large, resupinate; lip usually with a basal spur; column short or moderately long; the anther erect, basally firmly united with the column; two or four pollinia, sectile, with two basal, interlocular caudicles attached to two or one basal viscidia; stigma entire, concave.

DISTRIBUTION Primarily northern hemisphere, ranging into Africa and tropical Asia.

POLLINATION There is a relative wealth of information on the European species. Figure 8.7 is especially interesting, and indicates the importance

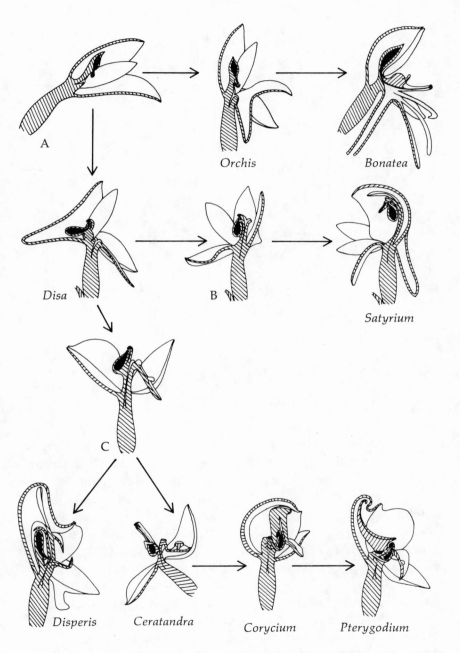

Figure 8.6 Diagrams showing probable relationships within the Orchideae and the Diseae. *(A)* Hypothetical ancestor. *(B,C)* Hypothetical intermediates. (After Vogel, 1959.)

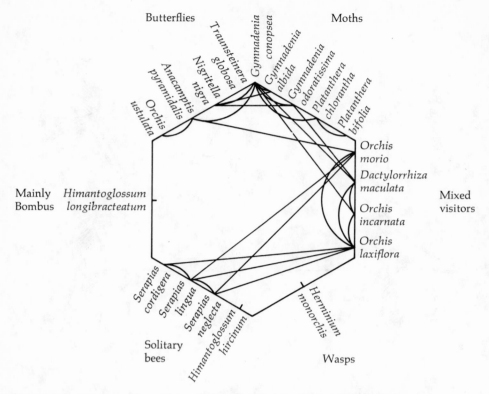

Figure 8.7 Diagram of crossing relationships between European Orchidinae, grouped according to their pollinators. Lines indicate reported hybrids. (After van der Pijl and Dodson, 1966.)

of pollinator classes in hybridization. The majority of the natural hybrids occur within a pollination class, or between species in the mixed visitor class and members of another class. As the figure indicates, this sub-tribe has a wide spectrum of pollinators, with bees, wasps, moths, and butterflies all being important. *Orchis* normally lacks nectar, with the notable exception of *O. coriophora*, which achieves a high percentage of pollination (Eberle, 1974; Peisl and Forster, 1975; Vöth, 1975). *Orchis papilionacea* on Elba is commonly pollinated by patrolling (territorial) males of *Eucera tuberculata* (Anthophoridae). Vogel found about 50 percent pollination, and suggests that such pollination may have played a role in the evolution of nectarless spurs. Such a system might also be the first step in the evolution of pseudocopulation, as found in *Ophrys*. *Platanthera* species are usually adapted for pollination by Lepidoptera, and *P. obtusata* is pollinated by mosquitoes in the northern part of its range. *Ophrys*, of course, is pollinated by male bees and wasps through pseudocopulation. Kullenberg and Bergström (1976) give a recent review of *Ophrys* pollination and attempt to correlate the species of *Ophrys* with the evolution of their pollinators.

CHROMOSOME NUMBERS   30, 32, 36, 38, 40, 42.

SPECIES   About 600.

GENERA   36 in four tentative alliances (see discussion): (1) Lacking tuberoids: *Amerorchis, Aorchis, Chondradenia, Galearis.* (2) tuberoids palmate or attenuate: *Brachycorythis, Chusua, Coeloglossum, Dactylorhiza, Gymnadenia, Hemipilia, Nigritella, Platanthera, Pseudodiphryllum, Pseudorchis, Schwartzkopffia, Silvorchis.* (3) Tuberoids spheroid: *Aceras, Amitostigma, Anacamptis, Barlia, Chamorchis, Comperia, Himantoglossum, Neobolusia, Neotinea, Neottianthe, Ophrys, Orchis, Piperia, Schizochilus, Serapias, Steveniella, Symphyosepalum, Traunsteinera.* (4) Hairy plants with flat, basal leaves, petals and lip often fimbriate, African: *Bartholina, Holothrix.*

INTERGENERIC HYBRIDS   This subtribe is outstanding in the great number of natural hybrids that are known (fig. 8.8). Vermeulen (1947) suggested

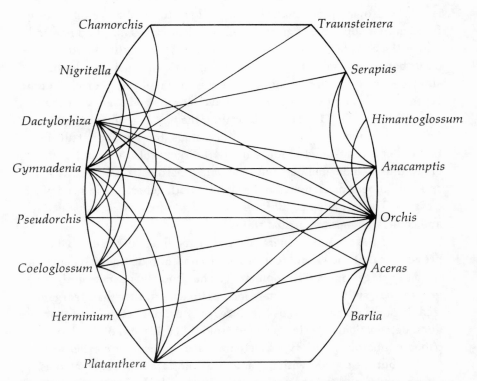

Figure 8.8   Known intergeneric hybrids in the Orchidinae. All of these are naturally occurring hybrids.

that all *Orchis–Dactylorhiza* hybrids were questionable. Other authors indicate that × *Orchidactyla* hybrids do occur but that they are relatively infrequent (see especially Peitz, 1972; Potůček, 1968). Unfortunately, the relative frequency of orchid hybrids is not easily obtained from the literature, most authors being content to indicate that a hybrid does or does not occur. Nevertheless, it seems clear that there are two alliances here, the *Orchis* alliance, including *Aceras, Anacamptis, Himantoglossum*, and *Serapias*, and the *Dactylorhiza* alliance, with *Coeloglossum, Gymnadenia, Nigritella, Platanthera*, and *Pseudorchis*. The members of the *Orchis* alliance are characterized by spheroid tuberoids, while the genera of the *Dactylorhiza* alliance all have palmate tuberoids (fig. 3.6). Hybrids between members of the two alliances do occur, but they are less frequent and less fertile. At least, the occurrence of back-crosses in × *Orchiaceras* and × *Gymnigritella* suggest that they are relatively fertile.

DISCUSSION   Though we have a great deal of information on many aspects of the European members of this group, a synthesis is badly needed. Nearly 40 genera and about 600 species are recognized, and it is easy to drown in the details but rather hard to see the overall pattern. Some authors divide this group into several subtribes, but I do not find any of the classifications entirely convincing. When natural hybrids occur between members of different subtribes, one must ask, do the subtribes really mean anything, or are they a classification of viscidium types rather than of whole plants? I have reorganized somewhat the groups of Senghas (1973–74), separating the first two alliances on features of the tuberoids rather than floral details, but I am not at all sure that either arrangement is adequate. Both *Brachycorythis* and *Hemipilia* are said to include members with palmate tuberoids and others with spheroid tuberoids. Senghas suggests that both genera are in need of revision. The last alliance, *Bartholina* and *Holothrix*, is very distinctive and could well merit subtribal status.

Some of the Eurasian genera seem to be too finely divided. My impression is that the wrong features were chosen to distinguish genera in the early stages of classification, and, as our knowledge increased, the genera were split up to eliminate inconsistencies, rather than reorganized. In some cases, vegetative features may be more significant than the floral details. Some species have spheroid tuberoids and others have palmate tuberoids. Both types were included in the older concepts of *Orchis*, but the species with palmate tuberoids are now segregated as *Dactylorhiza*, and the separation is well justified. However, one could unite *Aceras* and other genera with *Orchis,* and *Gymnadenia* with *Dactylorhiza.*

REFERENCES   Ackerman, 1977 *(Piperia);* Kullenberg and Bergström, 1976 (pollination of *Ophrys);* Senghas, 1973–74 (general); Senghas and Sundermann, 1968 *(Dactylorhiza),* 1972 *(Orchis);* Smith and Snow, 1976 (pollination of *Platanthera);* Stoutamire, 1974a (pollination of *Platanthera);* Summerhayes, 1956 (revision of *Brachycorythis);* Sundermann, 1964 *(Ophrys);* Vogel, 1972 (pollination of *Orchis papilionacea).*

Subtribe HABENARIINAE Bentham

DESCRIPTION   Habit: terrestrial, rarely epiphytic, with spheroid or oblong root-stem tuberoids; stems slender. Leaves: spiral, scattered or basal and clustered, convolute, conduplicate, nonarticulate. Inflorescence: terminal, simple, of one to many spiral flowers. Flowers: small to medium-large, usually resupinate; lip usually with a basal spur; column short; anther erect, basally firmly united with the column; two or four pollinia, sectile, with two basal, interlocular caudicles attached to two or one basal viscidia; the viscidia often borne on long rostellar stalks; stigma convex, entire or two-lobed, the lobes often stalked. (See fig. 8.9.)

DISTRIBUTION   Africa and pantropical, ranging into Eurasia.

POLLINATION   We have several records of moth pollination for *Habenaria,* and one of butterfly pollination for *Bonatea.* Pollination by wasps and beetles is reported for *Herminium.* Lepidopteran pollination is indicated for most of this subtribe.

CHROMOSOME NUMBERS   28, 32, 36, 40, 42.

SPECIES   About 1,100.

GENERA   21 in two tentative alliances: (1) Stigmas sessile or very short-stalked, entire or partly divided: *Androcorys(?), Benthamia, Diphylax, Gennaria, Herminium, Pecteilis, Peristylus, Smithorchis, Tylostigma.* (2) Stigmas 2, distinctly stalked: *Arnottia, Bonatea, Centrostigma, Cynorkis, Diplomeris, Habenaria, Megalorchis, Physoceras, Platycoryne, Roeperocharis, Stenoglottis, Tsaiorchis.*

INTERGENERIC HYBRIDS   The genus *Herminium* is known to hybridize with both *Aceras* and *Pseudorchis* (Orchidinae). One artificial hybrid between *Habenaria* and *Pecteilis* is registered.

DISCUSSION   This subtribe is notable for the development of rostellum arms (and long caudicles) and stalked stigmas. This is most prominent in *Bonatea* and some species of *Habenaria.* The placement of *Herminium* is rather problematical, and it may simply be a connecting link between the Habenariinae and the Orchidinae. Vermeulen (in Landwehr, 1977)

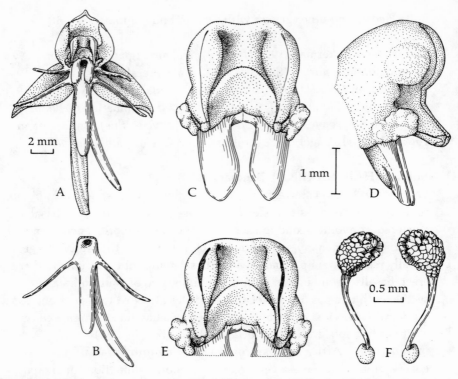

Figure 8.9 *Habenaria entomantha* (Orchidoideae: Orchideae). *(A)* Flower, front view. *(B)* Lip and Spur. *(C)* Column, front view. *(D)* Column, side view. *(E)* Column, after removal of hemipollinaria. *(F)* Hemipollinaria.

assigns *Herminium* to the *Platanthera* alliance, which fits well with the natural hybrids that are known, yet *Herminium* seems to be closely allied to *Peristylus*. The placement of *Androcorys* is even more problematical. Senghas suggests that it may be closely allied to *Herminium,* but the genus is, at best, poorly known.

REFERENCES   Seidenfaden, 1977 (Thai species); Senghas, 1973–74 (general).

Subtribe HUTTONAEINAE Schlechter

DESCRIPTION   Habit: terrestrial, with spheroid root-stem tuberoids; stems slender. Leaves: spiral, few, convolute, conduplicate, nonarticulate. Inflorescence: terminal, simple, of several spiral flowers. Flowers: medium, resupinate, sepals and petals fimbriate; lip without a spur; column short; anther erect, the locules together apically and diverging basally; two pollinia, sectile, with two basal, interlocular caudicles and two viscidia; stigma entire.

DISTRIBUTION.   Southern Africa.

POLLINATION   Not known.

CHROMOSOME NUMBER   Not known.

SPECIES   5.

GENUS   *Huttonaea.*

DISCUSSION   This subtribe is very distinctive in the form of the column and apparently has no close allies.

REFERENCE   Schelpe, 1966 (illustration and description).

Tribe DISEAE Dressler, 1979b

DISCUSSION   This tribe is primarily African, though *Disperis* and *Satyrium* each have representatives in tropical Asia. These subtribes are closely related to the *Orchideae,* but are distinguished by the anther bending back from the column. In some cases the anther is completely reversed, as compared to the Orchideae (see fig. 8.6).

PHYLETIC TRENDS   This group shows a complex pattern of evolution and is surely the most advanced of the primarily terrestrial groups. The common ancestor of the Diseae probably lacked a spur. The Disinae have developed a spur on the dorsal sepal, and we see all stages from a merely saccate sepal to a well-developed spur. In the Satyriinae we find paired spurs on the lip and, again, every stage from shallow, saccate nectaries to long spurs. In the Coryciinae, *Ceratandra,* in which the lip does not have a prominent appendage, would seem to be the least specialized of the subtribe, while the other three genera are all highly specialized, each in a different way. *Pterygodium* is noteworthy, as here the hood produced by the dorsal sepal and the petals forms a spur-like extension, the nectariferous tissue being supplied by an extension of the lip appendage. *Disperis* may show the same sort of spur, and at the same time has a small spur on each lateral sepal. (See fig. 8.10.)

Subtribe DISINAE Bentham

DESCRIPTION   Habit: terrestrial, with root-stem tuberoids; stems slender. Leaves: spiral, scattered, or basal and clustered, convolute, conduplicate, nonarticulate. Inflorescence: terminal, simple, of few to many spiral flowers. Flowers: small to medium-large, resupinate, dorsal sepal saccate or deeply spurred; lip usually very small or narrow, may be stalked or fringed; column very short, anther erect or usually inclined dorsally, basally completely united with column; two pollinia, sectile, with basal caudicles and one or two viscidia; stigma entire, very near base of perianth.

DISTRIBUTION   Africa, and especially southern Africa.

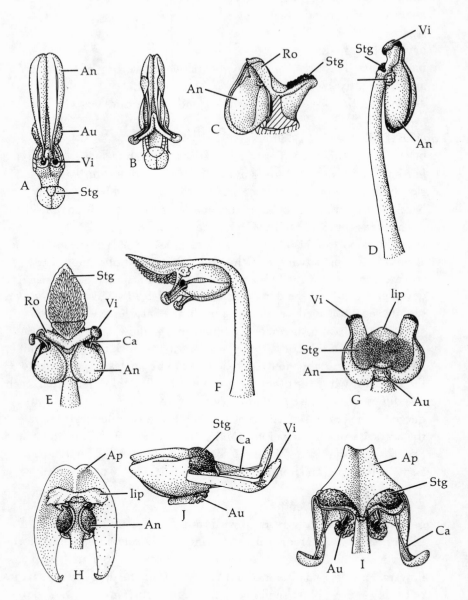

Figure 8.10   Columns of Diseae. *(A) Disa bivalvata. (B) Disa filicornis. (C) Disa draconis*, side view. *(D) Satyridium rostratum*, side view. *(E,F) Satyrium saxicolum*, front and side views. *(G) Ceratandra globosa. (H) Corycium oroban-choides. (I,J) Disperis paludosa*, front and side views. An, anther. Ap, appendage of the lip. Au, auricle. Ca, caudicle. Ro, rostellum. Stg, stigma. Vi, viscidium. (After Vogel, 1959.)

POLLINATION   We have only a few pollination records for *Disa*, indicating butterfly, bee fly, and fly pollination for different species. We also have Vogel's (1959) detailed analysis of the group, suggesting that the Disinae have radiated to a number of pollination systems.

CHROMOSOME NUMBER   Not known.

SPECIES   About 190.

GENERA   9: *Amphigena, Brownleea, Disa, Forficaria, Herschelia, Monadenia, Orthopenthea, Penthea, Schizodium.*

DISCUSSION   This group shows an impressive diversity of flower structure, based throughout on the development of a nectary by the dorsal sepal. The column structure is distinctive, the stigma being borne very near the base of the perianth, and the anther leaning back, in some cases being horizontal (perpendicular to the axis of the flower). The diversity of this group does not sort into genera as neatly as one might like, and some authors treat most of the genera listed here as synonyms of *Disa*. I follow Schelpe in the listing of genera.

REFERENCES.   Schelpe, 1971 (general); Senghas, 1973–74 (general); Vogel, 1959 (flower structure and function).

Subtribe SATYRIINAE Schlechter

DESCRIPTION   Habit: terrestrial, with root-stem tuberoids. Leaves: spiral, scattered or basal and more or less clustered, convolute, conduplicate, nonarticulate. Inflorescence: terminal, simple, of few to many spiral flowers. Flowers: small to medium, nonresupinate; lip with two saccate nectaries or spurs; column elongate, arched, the anther basally firmly united with the column, bent back so that the base is uppermost; two pollinia, sectile, with two basal interlocular caudicles and two viscidia; stigma somewhat two-lobed, somewhat overtopping the anther.

DISTRIBUTION   Primarily African, with *Satyrium* extending to tropical and nontropical Asia.

POLLINATION   The few available records indicate fly pollination, and *Satyrium pumilum* shows the syndrome of carrion fly pollination. However, some species have much longer spurs and are more likely to be pollinated by Lepidoptera or bees.

CHROMOSOME NUMBER   Not known.

SPECIES   About 110.

GENERA   3: *Pachites, Satyridium, Satyrium.*

DISCUSSION   Though related to the Disinae, this group seems clearly distinguished by the nonresupinate flowers, by the nectariferous lip with two depressions or spurs, and by the slender column with the anther reversed in its orientation.

REFERENCES   Senghas, 1973–1974 (general); Vogel, 1959 (flower structure and function).

Subtribe CORYCIINAE Bentham

DESCRIPTION   Habit: terrestrial or saprophytic, with root-stem tuberoids,
stem slender. Leaves: spiral, subbasal or much reduced, convolute, conduplicate, nonarticulate. Inflorescence: terminal, simple, of one to many
spiral flowers. Flowers: small to medium, resupinate or not; lateral sepals
often each with a small spur; petals often united to dorsal sepal; lip
united with column, usually very small, may bear an "appendage" which
is larger than the blade itself; column relatively short, anther basally
firmly united with column, bent backward so that the apex is toward
the base of the column, the anther thecae widely separated, each projecting forward with a viscidium at apex of the anther/rostellar arm;
two pollinia, sectile, each with a prominent interlocular caudicle and a
viscidium; stigma two-lobed, dorsal, well below the viscidia. (See fig.
8.11.)

DISTRIBUTION   Primarily African, with *Disperis* extending into tropical
Asia.

POLLINATION   The only observation for this remarkable group is the

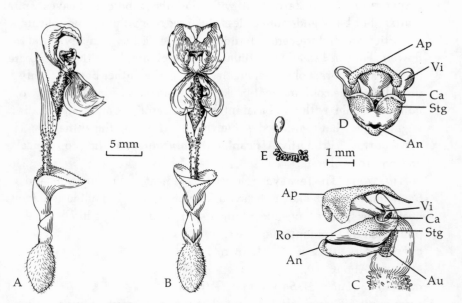

Figure 8.11   *Disperis pusilla* (Orchidoideae: Diseae). *(A)* Habit, showing flower in
side view. *(B)* Habit, front view. *(C)* Lip and column, side view. *(D)* Lip and
column, front view. *(E)* Hemipollinarium. An, anther. Ap, appendage of lip.
Ca, caudicles. Ro, rostellum. Stg, stigma. Vi, viscidium. (After Verdcourt, 1968.)

pollination of a *Disperis* species by a bee fly. From his study of the floral features, Vogel (1959) suggests that most are pollinated by Hymenoptera (probably bees), but that some may well be fly flowers.

CHROMOSOME NUMBER   Not known.

SPECIES   About 110.

GENERA   4: *Ceratandra, Corycium, Disperis, Pterygodium.*

DISCUSSION   This group is characterized by the upside-down anthers and a high degree of union between lip and column. The lip itself is usually very small, but in some cases it has an appendage that is larger than the lip. This group shows the most highly modified flowers of the subfamily (if not of the whole plant kingdom), and even with the help of Vogel's excellent study, trying to understand the homology of these flowers can give one a headache. Some species of *Disperis* are apparently somewhat saprophytic, and these seem to be reduced to a tuberoid, a bizarre flower larger than the tuberoid, and enough stem to connect the two (fig. 8.11).

REFERENCE   Vogel, 1959 (flower structure and function).

Anomalous Tribes

We have divided the orchids into six subfamilies, which seem to me to be natural groups. Unfortunately, there are a few leftovers that do not fit clearly into any of these six subfamilies. One could, at one extreme, shove them in somewhere and pretend that they fit, or, at the other extreme, name a new subfamily for each group of misfits. Neither of these courses of action seems very satisfactory, and so I will list them here, as misfits. When we know more about them, some of the problems may disappear. All of these leftovers fall into the category of terrestrials with soft pollinia, and so they are appended to this chapter.

Tribe TRIPHOREAE Dressler, 1979b

DESCRIPTION   Habit: terrestrial or saprophytic; stem slender, roots often fleshy or with nodular tuberoids. Leaves: subdistichous or solitary at midstem, or much reduced in near-saprophytes; convolute, pleated or conduplicate, nonarticulate. Inflorescence: terminal, simple, of one to several spiral flowers. Flowers: small, resupinate or not; column slender, anther erect, the thecae nearly as long as anther, or restricted to basal portion; pollen soft, mealy, as tetrads, with or without viscidium; stigma entire. (See fig. 8.12.)

DISTRIBUTION   Tropical America and into North America.

POLLINATION   Some species are self-pollinating; otherwise not known. The flower structure suggests pollination by small bees.

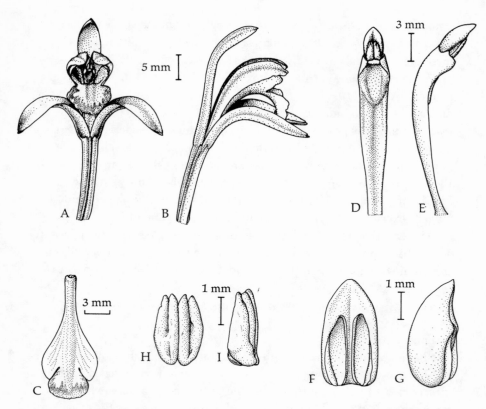

Figure 8.12 *Psilochilus mollis* (Triphoreae). *(A)* Flower, front view. *(B)* Flower, side view. *(C)* Lip, flattened. *(D)* Column, ventral view. *(E)* Column, side view. *(F)* Anther, ventral view. *(G)* Anther, side view. *(H)* Pollinia, upper surface. *(I)* Pollinia, side view.

CHROMOSOME NUMBER   44 rather small chromosomes.

SPECIES   About 20.

GENERA   3: *Monophyllorchis, Psilochilus, Triphora.*

DISCUSSION   These genera have been assigned to the Pogoniinae, but the erect anther of *Triphora* and *Psilochilus* just does not fit the epidendroid line of evolution (fig. 8.12). Many authors have included *Triphora* not only in the Pogoniinae but in *Pogonia* itself. However, an adequate series of differences are shown by Ames (1922) and Baldwin and Speese (1957). The three genera listed here give a nice sequence from the fleshy, clearly erect anther of *Triphora* to the nearly operculate anther of *Monophyllorchis,* in which the thecae are limited to the basal portion of the anther. The habit of *Psilochilus* is unusual; the plants look, at first glance, like Commelinaceae, and the manner of growth rather simulates that of the

Goodyerinae. All three genera have definite (perigenous) subsidiary cells. *Psilochilus* and *Triphora*, or at least those that are not self-pollinating, show gregarious flowering.

RELATIONSHIPS   On general features, this group agrees well with the Neottieae and the more primitive Diurideae, but it does not seem to be very closely related to these or to any other group. The trapezoid subsidiary cells suggest a closer alliance with the Epidendroideae and the Vandoideae, and the anther of *Monophyllorchis* is suggestive of the vandoid anther, but there is a very large gap between this group and the Vandoideae. The Triphoreae appear to be a relic group, with no very close allies. On present evidence, this small group could be placed in the Orchidoideae, and this is supported by current (unpublished) work on pollen and seed structure.

REFERENCES   Ames, 1922 (comparison of *Triphora* and *Pogonia*); Baldwin and Speese, 1957 (chromosomes); Sweet, 1969 (*Monophyllorchis*).

Tribe WULLSCHLAEGELIEAE Dressler, 1980a

DESCRIPTION   Habit: saprophytic, stems slender, roots fusiform, leafless, aerial parts clothed with bifurcate hairs. Inflorescence: terminal, of many spiral flowers, the apex nodding in bud. Flowers: tiny, resupinate or not, sepals free, lip simple, basally saccate; column short, the anther erect, very fleshy, embedded in the column, with curved thecae; two pollinia, sectile, with very slender massulae, stigma entire, on apex of column, concave, with a prominent viscidium that is attached to the pollinia ventrally. (See fig. 8.13.)

DISTRIBUTION   Tropical America.

POLLINATION   Self-pollinated.

CHROMOSOME NUMBER   Not known.

SPECIES   2.

GENUS   *Wullschlaegelia*.

DISCUSSION   I have been reluctant to create a subfamily for this insignificant genus (it would ruin the symmetry of my phyletic diagrams), but I am unable to associate this genus with any of the named subfamilies. This genus has been placed in the Cranichidinae for many years, but the flowers are so tiny that no one paid much attention to their details. The flowers are nonresupinate in one species (which fits the Cranichidinae) and resupinate in the other (which does not). The short column with the anther surrounded by columnar tissue is unusual. The anther is very fleshy, and the thecae are markedly curved. The pollinia are sectile, with very narrow massulae, without caudicles but with a prominent viscidium. As the flowers are self-pollinating, the viscidium is

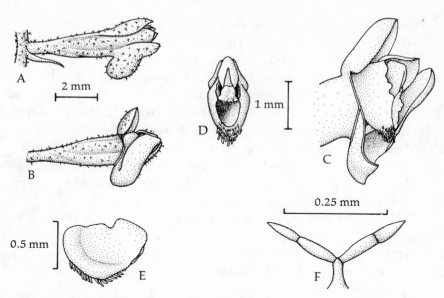

Figure 8.13 *Wullschlaegelia calcarata* (Wullschlaegelieae). *(A)* Flower, side view. *(B)* Flower, side view, with lateral sepal removed. *(C)* Column, side view, with sepals and petals in place. *(D)* Column, front view. *(E)* Anther, side view. *(F)* Branched hair from inflorescence.

nonfunctional. Aside from their distinctive flower structure, they have spindlelike roots, unlike those of any other orchid known to me.

RELATIONSHIPS   The nodding inflorescence of this orchid is reminiscent of *Epipogium*, which is also saprophytic, but in other respects the two genera are very different. I cannot indicate any clear relationships between *Wullschlaegelia* and any other orchid group, though more information on the Epipogieae, and especially *Stereosandra*, may indicate a definite link between this tribe and *Wullschlaegelia*.

REFERENCE   Dressler, 1980 a (general).

The Epidendroid Orchids

9

We tend to think of the four subfamilies treated in chapters 7 and 8 as the more primitive groups, but in fact they include some very highly evolved orchids. In many respects, the more primitive epidendroids have some features more primitive than we find in any member of the Spiranthoideae or Orchidoideae—especially the habit, fruits, and seeds of the Vanillinae. In terms of number of species, the Epidendroideae are *the* major orchid group, with more than half of all orchid species in this one subfamily. Naturally, such a large group is rather diverse.

Subfamily

EPIDENDROIDEAE Lindley

DISCUSSION   Most of the orchids in this subfamily have hard, clearly defined pollinia, but there is no clear line between the primitive members with soft, mealy pollen and the more advanced members with clearly hard pollinia. The one feature which ties together most of this group is to be found in the development of the anther. In most Epidendroideae, and in all the primitive genera, the anther is erect in the young flower bud, but as the bud develops, the anther bends down over the apex of the column until it is at a right angle to the axis of the column, or often even markedly ventral. This was shown by Hirmer (1920), through microscope sections of the flower buds, but one can see it quite clearly by dissecting the buds of different ages (fig. 3.20). In the primitive Epidendroideae this "incumbent" anther functions very nicely by swiveling on its point of attachment to place pollen on an insect as it leaves the flower (fig. 3.21). There are many groups in the subfamily that have lost the primitive orientation of the anther, and these are largely the groups that have developed viscidia. In most cases, though, their relationship to the other Epidendroideae is very clear. Since the anther is erect in the early stages of flower development, it is very "easy" for it to retain this position, once the development of a viscidium renders the primitive swivel mechanism unnecessary. I suggest that the efficiency of the swiveling incumbent anther has been important in permitting the survival of very primitive orchids in this group.

RELATIONSHIPS   The Epidendroideae and the Vandoideae are the two subfamilies in which tubers, corms, and pseudobulbs are to be found, and they have similar subsidiary cells. Thus, it is quite possible that they share a close common ancestry, but the epidendroid line clearly diverged from other lines early in orchid evolution. The Epidendroideae also show relationship to the Cypripedioideae and may be about equally related to the Orchidoideae.

Tribe VANILLEAE Blume

DISCUSSION   The Vanilleae are characterized by soft, mealy pollen masses that are not at all divided, and usually by a considerable degree of union between the lip and the column. We find within this group a number of relatively primitive features. The transoceanic distributions of *Epistephium, Vanilla,* and the Palmorchidinae also suggest an old group.

RELATIONSHIPS   The living Vanilleae are not, of course, ancestral to any other orchid group, but they give us an idea of what such an ancestral group might have been like. The Vanilleae show a definite relationship to the Arethuseae, and we may guess that all the other tribes of this subfamily have evolved from something like the Vanilleae.

PHYLETIC TRENDS   The genus *Vanilla* is off by itself in habit, and *Galeola* is somewhat more specialized as a saprophyte. In general features, *Epistephium* and *Palmorchis* are probably closer to the primitive condition for this tribe. The Pogoniinae and *Lecanorchis* are smaller, herbaceous plants but still have much the same flower structure.

Subtribe VANILLINAE Lindley

DESCRIPTION   Habit: shrubs or vines, sometimes saprophytic; stem woody or fleshy; roots not thickened. Leaves: spiral or subdistichous, sometimes reduced to scales, convolute, conduplicate and fleshy, or leathery and more or less net-veined, nonarticulate. Inflorescence: terminal or axillary, simple, of few to many spiral flowers. Flowers: small to large, resupinate, usually with an abscission layer between ovary and perianth, lip usually partly united with the column, often more or less trumpet-shaped; column slender, anther terminal, incumbent, generally somewhat ventral; pollen soft and mealy, as monads; stigma entire or three-lobed, usually emergent, without viscidium. Fruit: one-locular or three-locular, capsular or fleshy, often fleshy and opening as two unequal valves; seeds small, with a hard seed coat, may have prominent wings. (See fig. 9.1.)

DISTRIBUTION   Pantropical.

POLLINATION   Most articles about *Vanilla* tell of pollination by *Melipona beechei*, but authors have copied each other until the source seems hopelessly lost. In the American tropics, *Vanilla* is frequently pollinated by large bees of the genus *Eulaema*, and I doubt that a small bee such as *Melipona* could be an effective pollinator. *Eulaema* specimens occasionally have a flat, triangular pad of pollen on the scutellum, which I believe to be the pollen of *Vanilla*. The Euglossini do not occur in the Old World tropics, but other large bees are the most probable pollinators there and may play a role in tropical America, as well. Nothing is known of polli-

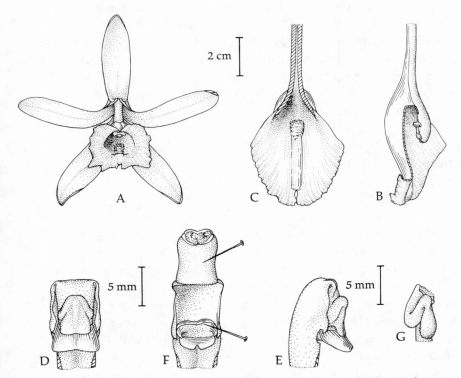

Figure 9.1   *Vanilla pompona* (Epidendroideae: Vanilleae). (*A*) Flower, front view. (*B*) Lip and Column, side view. (*C*) Lip, spread. (*D*) Apex of column, with anther in place. (*E*) Apex of column, side view. (*F*) Apex of column, with both anther and midlobe of stigma tipped back. (*G*) Anther, side view.

nation in the other genera, but bee pollination is indicated by flower structure.

CHROMOSOME NUMBERS   28, 32.

SPECIES   About 165.

GENERA   5: *Clematepistephium, Epistephium, Eriaxis, Galeola, Vanilla.*

DISCUSSION   In habit and seed structure these genera suggest something approaching the Ur-orchid, at least for this subfamily. *Epistephium* and *Eriaxis* show what may be a very primitive habit of growth. Whether the net-veined leaves are a primitive feature found only here, or a new development, it is difficult to say. *Vanilla* are usually vines, as is *Clematepistephium,* surely a specialization. *Galeola* are usually saprophytes and include much the largest saprophytes in the family.

REFERENCES   Bouriquet, 1954 *(Vanilla);* Garay, 1961 *(Epistephium).*

Subtribe LECANORCHIDINAE Dressler, 1979b

DESCRIPTION   Habit: saprophytic, leafless, stem slender, from a scaly rhizome. Inflorescence: terminal, simple, of few to several flowers. Flowers: small or medium small, resupinate or not(?), with a prominent "calyculus" and an abscission layer at base of perianth, sepals and petals free, similar, lip basally united with column, simple or three-lobed, the midlobe often hairy; column slender, anther incumbent, pollen soft and mealy, as monads; stigma three-lobed, without viscidium. Fruit: three-locular, capsular, seeds minute.

DISTRIBUTION   Tropical Asia.

POLLINATION   Not known; probably pollinated by bees or self-pollinating.

CHROMOSOME NUMBER   Not known.

SPECIES   About 20.

GENUS   *Lecanorchis.*

DISCUSSION   This genus of inconspicuous saprophytes has been tossed about from one group to another without fitting very well in any of them. Ackerman and Williams (in press a), in studying its pollen, find a strong indication of a relationship with *Vanilla.* The habit, and especially the minute seeds, are rather unlike the Vanillinae, but the union of lip and column and the calyculus like that of *Epistephium* suggest a place near the Vanillinae. A separate subtribe within the Vanilleae seems best.

Subtribe PALMORCHIDINAE Dressler, 1979b

DESCRIPTION   Habit: terrestrial, stem tough and reedlike, up to one meter tall. Leaves: spiral or subdistichous, convolute, plicate, nonarticulate.

Inflorescence: terminal, simple, of few to many spiral flowers. Flowers: small, resupinate; lip partly united with column; column slender, anther terminal, incumbent; four pollinia, soft but coherent; stigma entire, emergent, without viscidium.

DISTRIBUTION   Tropical America and tropical Africa.

POLLINATION   Not known; probably by small bees.

CHROMOSOME NUMBER   Not known.

SPECIES   About 12.

GENERA   2: *Diceratostele, Palmorchis.*

DISCUSSION   The creation of a separate subtribe for *Palmorchis* was suggested by Schweinfurth and Correll (1940), but, unfortunately they did not comply with the rules for the description of new taxa. Several authors have placed these genera with the Sobraliinae, but the four undivided pollinia do not fit well in that group. Further, the nonarticulate leaves and the overall flower structure are very similar to *Vanilla* and *Lecanorchis.* This seems much the best place for this small group.

REFERENCE   Schweinfurth and Correll, 1940 (revision of *Palmorchis*).

Subtribe POGONIINAE Pfitzer

DESCRIPTION   Habit: terrestrial or saprophytic; may have nodular root tuberoids or produce shoots from a rhizome-like root. Leaves: distichous or spiral (?), scattered, basal and clustered or whorled at midstem, convolute, conduplicate, nonarticulate. Inflorescence: terminal, simple, of one to several spiral flowers. Flowers: medium-small to large, with an abscission layer between ovary and perianth; column slender, anther terminal, incumbent; pollen soft and mealy, as monads or tetrads, without viscidia; stigma entire.

DISTRIBUTION   Tropical America, North America, and eastern Asia.

POLLINATION   Members of *Pogonia* deceive bees by the pollenlike cluster of hairs on the lip (Thien and Marcks, 1972); the other genera have gullet flowers and suggest bee pollination.

CHROMOSOME NUMBER   18 large chromosomes.

SPECIES   About 40.

GENERA   5: *Cleistes, Duckeella, Isotria, Pogonia, Pogoniopsis.*

DISCUSSION   Some authors have assigned all these genera to the Vanillinae, but they differ in a number of features. I follow Dunsterville and Garay (1976) in referring *Duckeella* to this subtribe.

REFERENCES   Ames, 1922 (North American members); Baldwin and Speese, 1957 (chromosomes); Thien and Marcks, 1972 (pollination of *Pogonia*).

Tribe GASTRODIEAE Lindley

DISCUSSION   This group, even more than the preceding tribe, is largely saprophytic. The pollinia are always sectile but without caudicles. As is usual for saprophytes, we do not have enough information on the details of flower structure.

RELATIONSHIPS   This saprophytic group seems to be an offshoot of the very early epidendroid line, and more related to the Vanilleae then to any other group.

PHYETIC TRENDS   The genus *Nervilia* has green leaves and is photosynthetic, but the rest of this group are saprophytic. The two most notable trends are to greater union of the perianth parts and going wholly underground in the Australian Rhizanthellinae.

Subtribe NERVILIINAE Schlechter

DESCRIPTION   Habit: terrestrial, with a spheroid corm. Leaves: only one, subcircular, basally truncate to deeply cordate, convolute, plicate, nonarticulate. Inflorescence: terminal, simple, of one to several spiral flowers. Flowers: medium-small, resupinate or erect; column slender, anther incumbent; two pollinia, sectile, without caudicles or viscidia; stigma entire.

DISTRIBUTION   Tropical Asia and Africa.

POLLINATION   Not known, but the flowers appear to be adapted to pollination by small bees.

CHROMOSOME NUMBERS   54, 72, 108, 144.

SPECIES   About 80.

GENUS   *Nervilia.*

DISCUSSION   This genus was, for a long time, treated as a synonym of *Pogonia.* The flowers are superficially similar, but the habit is very different, and the pollinia are sectile.

Subtribe GASTRODIINAE Lindley

DESCRIPTION   Habit: saprophytic, with fleshy tuber or coralloid underground stem (both may occur on same plant); leafless. Inflorescence: terminal, simple, of one to many spiral flowers. Flowers: small or medium-small, fleshy, resupinate or not(?), sepals and petals more or less united; column slender, often winged, usually with a prominent foot; anther terminal, incumbent; two pollinia, sectile, with or without a viscidium; stigma entire, sometimes at base of column. (See fig. 9.2.)

DISTRIBUTION   Pantropical, but primarily in tropical Asia and Australasia.

POLLINATION   Not known.

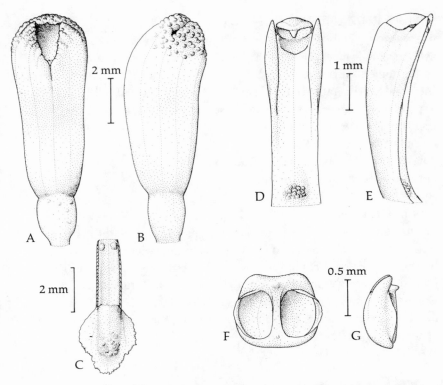

Figure 9.2 *Gastrodia siamensis* (Epidendroideae: Gastrodieae). *(A)* Flower, front view. *(B)* Flower, side view. *(C)* Lip, flattened. *(D)* Column, ventral view. *(E)* Column, side view. *(F)* Anther, ventral view. *(G)* Anther, side view. Drawn from material preserved in liquid.

CHROMOSOME NUMBERS   16, 18, ±36, 40, ±150.

SPECIES   About 50.

GENERA   6: *Auxopus, Didymoplexiella, Didymoplexis, Gastrodia, Neo-clemensia, Uleiorchis.*

DISCUSSION   There is one species of this group in tropical America, and a few in Africa, but the group is primarily Asiatic. The mycorrhizal relationships of the Gastrodieae have been carefully studied by Burgeff (1932), but other aspects of their biology are little known. All are saprophytic, and the sepals and petals are more or less united, giving them an aspect unusual among the primitive orchids. The pollinia are always sectile, as far as I can determine. Careful study of adequate material may show that some of the plants assigned here are not at all related to *Gastrodia*.

REFERENCE   Burgeff, 1932 (saprophytism and mycorrhizae).

Subtribe RHIZANTHELLINAE Rogers

DESCRIPTION   Habit: saprophytic, subterranean or barely reaching the soil surface; leafless, stem fleshy. Inflorescence: dense, headlike, of many spiral flowers, surrounded by large bracts. Flowers: small, fleshy, erect, the sepals and petals more or less united; column small, with small appendages; anther terminal, incumbent; pollen soft and mealy, sectile(?); stigma entire.

DISTRIBUTION   Australia.

POLLINATION   Probably self-pollinating.

CHROMOSOME NUMBER   Not known.

SPECIES   2.

GENERA   2: *Cryptanthemis, Rhizanthella*.

DISCUSSION   These subterranean orchids are the subject of many articles, but they are rarely seen, and not much is known about them. At one time I fancied that I could see a resemblance to the members of the Diurideae (Dressler and Dodson, 1960), but now I must admit that Rogers' original opinion—that they are close to the Gastrodiinae—is much more reasonable, especially considering the almost complete union of the sepals and petals in *Rhizanthella*.

REFERENCE   Hunt, 1953 (general).

Tribe EPIPOGIEAE Parlatore

DESCRIPTION   Habit: saprophytic from a coralloid underground stem or a fleshy tuber, leafless. Inflorescence: often thick and fleshy basally, simple, of few to many spiral flowers. Flowers: small to medium-large, nonresupinate or resupinate; lip simple or concave with a prominent basal spur; column short, anther fleshy, incumbent or more or less erect; two pollinia, sectile, with two caudicles, and a viscidium; stigma near base of column, entire. (See fig. 9.3.)

DISTRIBUTION   Eurasia, tropical Africa, and tropical Asia.

POLLINATION   Bumblebee pollination has been reported for *Epipogium aphyllum*, but van der Pijl and Dodson (1966) are doubtful. Most other species are self-pollinating.

CHROMOSOME NUMBER   68.

SPECIES   4 or 5.

GENERA   2: *Epipogium, Stereosandra*.

DISCUSSION   This tribe is an obvious relic, with four or five species in two genera. They are all saprophytes and are characterized by sectile pollinia with caudicles. Dressler and Dodson (1960), misled by a superficial resemblance, assigned *Epipogium* to the Orchideae. Vermeulen (1965) then made a careful study of the genus and showed convincingly

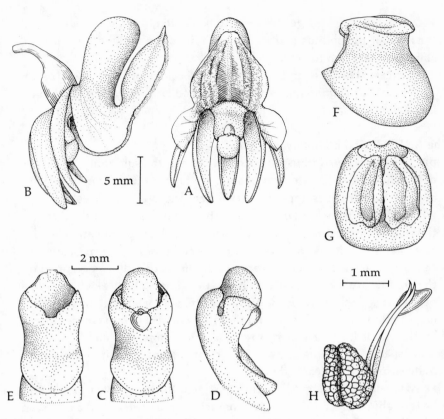

Figure 9.3 *Epipogium aphyllum* (Epidendroideae: Epipogieae). (*A*) Flower, front view. (*B*) Flower, side view. (*C*) Column, ventral view. (*D*) Column, side view. (*E*) Column, after removal of anther. (*F*) Anther, side view. (*G*) Anther, ventral view. (*H*) Pollinarium. (*A–G*) Drawn from material preserved in liquid.

that it could not be placed in the Orchideae. At the same time, he showed that it is extremely different from the Neottieae and from most other orchid groups with soft pollinia. The anther of *Epipogium* is somewhat incumbent, but unlike that of any other orchid.

Stereosandra has been placed with *Epipogium,* but the structure of the column, anther, and pollinia, and the spurless lip are all different. Vermeulen (1966) reports that it has a viscidium derived from the anther. The viscidium is not functional, of course, in a self-pollinated flower, but no other orchid is known with such a viscidium. I have not seen good material, but Brieger's suggestion that it should be placed in a separate subtribe has considerable merit. The anther is erect, but this may be a secondary condition in a self-pollinating population.

RELATIONSHIPS   These genera seem to be related to the Gastrodieae, but
the form of the pollinia and anther are quite different, and the relationship does not seem to be very close.

REFERENCES   Docters van Leeuwen, 1937 (ecology of *Epipogium roseum*);
Tuyama, 1967 *(Epipogium roseum)*; Vermeulen, 1965 (classification of
Epipogium).

Tribe ARETHUSEAE Lindley

DISCUSSION   The Arethuseae are a key tribe in the subfamily Epidendroideae. They show definite relationships to the primitive tribe Vanilleae,
and most other tribes appear to have been derived from ancestors
which, if they were known, would probably be classified in the Arethuseae. Most Arethuseae have corms, although *Arundina* and *Dilochia*
have slender stems with somewhat thickened bases. The leaves are plicate, or occasionally conduplicate, the stigma is usually emergent, and
the pollinia are soft or relatively hard, and almost always eight in number. That is, each theca of the anther is divided by a partition so that
eight pollinia are developed, and these are connected by ventral, intralocular caudicles. The genera *Arethusa*, and *Calopogon*, and
some species of *Bletilla*, are atypical in this regard as their pollinia are
quite soft and not clearly divided.

RELATIONSHIPS   This group is clearly related to the Vanilleae, and may
be thought of as having evolved from rather *Epistephium*-like ancestors.
The Coelogyneae and the Malaxideae may well be derived from something very close to the living Arethuseae. The Epidendreae seem a bit
more isolated, though they too could be derived from something like
Arundina.

PHYLETIC TRENDS   This group shows less diversity than many other
orchid tribes of comparable size. Most Sobraliinae have slender, reedlike
stems, as do *Arundina* and *Dilochia*. The rest of the genera are mostly
cormous and most have lateral inflorescences. There is relatively little
specialization in flower structure, though *Elleanthus* shows adaptation
to bird pollination in most species and *Calanthe* suggests adaptation to
Lepidopteran pollination. Both *Arethusa* and *Calopogon* attract pollinators by deceit and offer no real reward. Their soft, mealy pollinia may be
a sort of regression, related to pollination by hairy bumblebees, rather
than a primitive feature (Stoutamire, 1971).

Subtribe ARETHUSINAE Lindley

DESCRIPTION   Habit: terrestrial with a fleshy corm. Leaf: convolute,
plicate, nonarticulate. Inflorescence: terminal, simple, of one or few
spiral flowers. Flowers: medium-large, resupinate, lip arched, with

yellow, pollenlike hairs; a cavity at the base of the lip seems to lack nectar; column slender, arched, flattened and petaloid apically; anther ventral, incumbent; four pollinia, soft and mealy, sectile; stigma entire, emergent, without a viscidium.

DISTRIBUTION   North America.

POLLINATION   The genus *Arethusa* is pollinated by bumblebee queens in early spring, before they have learned which flowers have rewards and which do not (Stoutamire, 1971; Thien and Marcks, 1972).

CHROMOSOME NUMBER   Not known.

SPECIES   1.

GENUS   *Arethusa*.

DISCUSSION   Most classifications group *Arethusa* with a few other genera that have soft pollinia, but none of them is very close to *Arethusa*, and all of them seem closer to the Bletiinae.

REFERENCES   Stoutamire, 1971 (pollination); Tan, 1969 (comparison with *Bletilla* and *Calopogon*); Thien and Marcks, 1972 (pollination).

Subtribe THUNIINAE Schlechter

DESCRIPTION   Habit: terrestrial, with thick, fleshy stems. Leaves: distichous, duplicate, articulate(?). Inflorescence: terminal, of several spiral flowers, simple, pendant. Flowers: large, resupinate, lip trumpet-shaped, parallel with and enfolding the column; column slender, somewhat petaloid terminally, anther ventral, incumbent, four-celled; four pollinia(?), soft and mealy; stigma entire, emergent.

DISTRIBUTION   Tropical Asia.

POLLINATION   Not known; the flower structure suggests bee pollination.

CHROMOSOME NUMBER   42.

SPECIES   About 6.

GENUS   *Thunia*.

DISCUSSION   In the past, quite diverse genera with slender stems have been grouped with *Thunia* (which actually has a fleshy stem), but none of them is very similar in details. *Claderia* fits the Cymbidieae in its floral features, and *Arundina* and *Dilochia* are closer to Bletiinae than to *Thunia*. *Thunia* is unusual in the soft, elongate pollinia, which are different from those of allied subtribes in form, in the fleshy stem–pseudobulb, and in the conduplicate, glaucous leaves. It is adapted to a strongly seasonal climate and can survive extreme drought during the resting period.

Subtribe BLETIINAE Bentham

DESCRIPTION   Habit: terrestrial, epiphytic or saprophytic, with slender stems or usually with corms or pseudobulbs of several internodes.

Leaves: distichous, convolute, plicate or occasionally conduplicate, usually articulate. Inflorescence: lateral or sometimes terminal, simple, with several to many spiral flowers. Flowers: medium-small to large, resupinate or nonresupinate, lip sometimes saccate or with a prominent spur; column short or long, often with a prominent column foot; anther terminal, incumbent, usually eight-celled; eight pollinia or rarely four, hard or relatively soft, laterally flattened or clavate, with ventral or terminal, intralocular caudicles; stigma entire, emergent or not, with a viscidium in some genera. (See fig. 9.4.)

DISTRIBUTION   Pantropical, extending into warmer areas of China and Japan and into North America.

POLLINATION   Pollination by bees is reported for *Bletia*, *Arundina*, and *Phaius*, and to be expected for most genera. The flowers of *Calopogon*, like those of *Arethusa*, depend on naive bees for pollination. In this

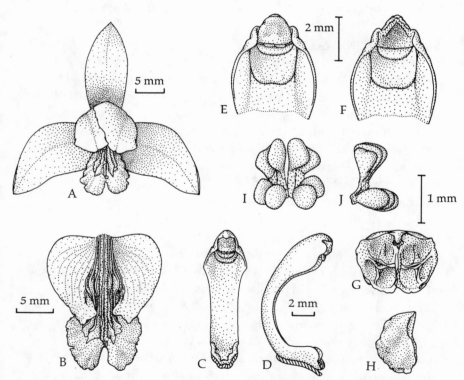

Figure 9.4   *Bletia purpurea* (Epidendroideae: Arethuseae). *(A)* Flower, front view. *(B)* Lip, flattened. *(C)* Column, ventral view. *(D)* Column, side view. *(E)* Apex of column, with anther in place. *(F)* Apex of column, after removal of anther. *(G)* Anther, ventral view. *(H)* Anther, side view. *(I)* Pollinarium, top view. *(J)* Pollinarium, side view.

case, the lip is hinged, and the weight of the pollinator is sufficient to tip it over and throw the insect against the column. *Calopogon tuberosus*, the species with largest flowers, is usually pollinated by queen bumblebees and carpenter bees, while other species are pollinated by smaller bees, such as halictine bees. Fly pollination is reported for *Plocoglottis*, and Lepidopteran pollination is to be expected in *Calanthe*.

CHROMOSOME NUMBERS   32, 36, 38, 40, 42, 44, 48, 50, 58.

SPECIES   About 380.

GENERA   26 in seven tentative alliances: (1) *Arundina, Dilochia*. (2) *Calopogon*. (3) *Acanthephippium, Ancistrochilus, Anthogonium, Aulostylis, Bletia, Bletilla, Calanthe, Cephalantheropsis, Eleorchis, Gastrorchis, Hexalectris, Ipsea, Pachystoma, Phaius, Spathoglottis*. (4) *Plocoglottis*. (5) *Hancockia, Mischobulbon, Nephelaphyllum, Tainia, Tainiopsis*(?). (6) *Coelia*. (7) *Chysis*.

INTERGENERIC HYBRIDS   Artificial hybrids are registered between *Calanthe* and *Gastrorchis* between *Phaius* and *Gastrorchis*, and between *Phaius* and *Calanthe*. Tanaka (1971) reports crosses between *Arundina* and *Bletilla* and between *Bletilla* and *Eleorchis*.

DISCUSSION   The continued transfer of misfits from other groups has enlarged the Bletiinae considerably, but its diversity has not been increased proportionately. Still, further study may show that some of the above alliances deserve to be placed in their own subtribes. *Bletilla, Eleorchis* and *Calopogon* have soft pollinia, but are otherwise like the Bletiinae (Tan, 1969). *Bletilla formosana* is illustrated as having typical *Bletia* pollinia (*Journal of Geobotany*, 14:23, fig. 115). Both *Arundina* and *Dilochia* have slender stems and conduplicate leaves, but their floral features agree well with the Bletiinae, and both slender stems and terminal florescences occur in other members of this subtribe. *Plocoglottis* is perhaps the most aberrant genus in the group as it is treated here. It has only four pollinia with large caudicles and a well-developed viscidium. The fifth series in the list was formerly included with *Diglyphosa* in a separate subtribe, but the pollinia of that supposed subtribe were of two very different sorts. Some of this group have a distinctive habit, with one-leaved pseudobulbs alternating with leafless, reduced pseudobulbs that bear terminal inflorescences, but some species of *Tainia* seem to be typical of the Bletiinae in their habit. *Coelia* is unusual in its spherical pseudobulbs of a single internode and the more or less conduplicate leaves, but it fits better here than in any other subtribe (Pridgeon, 1978). Finally, *Chysis* shows an unusual habit, with clublike pseudobulbs, and the pollinia are distinctive. Schlechter (1926) treated it as a separate subtribe, the Chysinae. *Hexalectris* has been grouped with the Corallorhizinae, but its flower structure does not fit there and is at home in the Bletiinae.

Calanthe has clavate pollinia with a distinct viscidium, and the lip is often united with the column. These features, with the prominent spur, suggest Lepidopteran pollination for the genus. Its features seem to be quite analogous with those of Epidendrum.

REFERENCES   Pridgeon, 1978 (Coelia); Seidenfaden, 1975a (Calanthe in Thailand); Tan, 1969 (Bletilla and Calopogon); Teoh and Lim, 1978 (chromosomes); Thien, 1973; Thien and Marcks, 1972 (pollination of Calopogon).

Subtribe SOBRALIINAE Schlechter

DESCRIPTION   Habit: terrestrial or epiphytic, stems slender, usually elongate. Leaves: distichous or subdistichous, usually convolute and plicate, but duplicate in some cases, fleshy in Arpophyllum, articulate. Inflorescence: terminal or lateral, usually of few to many spiral or distichous flowers. Flowers: small to large, membranous, resupinate or nonresupinate; lip more or less enfolding the column, often trumpet-shaped, simple, usually with prominent calluses, sometimes basally saccate; column short or elongate, often with armlike wings; anther terminal, two- to eight-celled; eight soft pollinia, superposed or ovoid; stigma usually emergent, with a distinct viscidium in some cases.

DISTRIBUTION   Tropical America.

POLLINATION   Most Sobralia species are pollinated by various kinds of bees, while a few show obvious adaptations to hummingbird pollination. Arpophyllum, Elleanthus, and Sertifera all show the hummingbird syndrome, and hummingbirds have been seen visiting Elleanthus on several occasions.

CHROMOSOME NUMBER   Not known.

SPECIES   About 150.

GENERA   5: Arpophyllum, Elleanthus, Sertifera, Sobralia, Xerorchis.

DISCUSSION   This is the most distinctive subtribe of the Arethuseae, and may be only remotely related to the Bletiinae. The stems are thin and woody in this group. The pollinia of Sobralia are usually distinctive, the basal (with respect to the anther) four pollinia being superposed, and the distal four being twisted together and acting as caudicles. In the hummingbird-pollinated species, however, the pollinia are ovoid like those of Elleanthus. Arpophyllum is distinctive in its habit, but the flower structure is like that of Elleanthus, and it is less of a misfit here than in the Laeliinae. Xerorchis has been placed in the Vanillinae by some, but the eight pollinia are discordant there. The thin, membranous flowers of this subtribe make poor herbarium specimens, and their taxonomy is, as one would expect, poorly understood.

REFERENCES   Garay, 1974a (revision of *Arpophyllum*).

Tribe COELOGYNEAE Pfitzer

DISCUSSION   In previous classifications (including my own) this group usually has been placed with the Epidendreae, but it is very distinct in its features and shows greater affinity with the Arethuseae. The pseudo-bulbs are each made up of a single internode, the pollinia are super-posed or ovoid with prominent, massive caudicles. The column is often more or less petaloid, and the stigma is often emergent.

RELATIONSHIPS   I believe that the Coelogyneae are most closely related to the Arethuseae and probably derived from Arethuseae-like an-cestors. The pollinia of *Coelogyne* could easily evolve from something like those of most *Sobralia* species.

PHYLETIC TRENDS   I would guess that the smaller-flowered *Coelogyne* species approach the primitive condition for the tribe, and that other genera have evolved through reduction of flower size, aggregation of flowers, the development of ovoid or somewhat clavate pollinia, and the union of lip with column. *Otochilus*, which forms new pseudobulbs from the apices of older ones, simulates the habit of the American *Scaphy-glottis*.

Subtribe COELOGYNINAE Bentham

DESCRIPTION   Habit: epiphytic or terrestrial, with pseudobulbs or corms of a single internode. Leaves: convolute or duplicate, plicate or conduplicate, articulate. Inflorescence: terminal, often produced before growth of pseudobulbs, simple, of few to many flowers, spiral or distichous. Flowers: small to large, resupinate; base of lip may be saccate; column short or elongate, apex often petaloid and hooded over anther; anther terminal or ventral, incumbent; four or two pollinia, these superposed or ovoid, with prominent caudicles; stigma entire, often emergent. (See fig. 9.5.)

DISTRIBUTION   Tropical Asia and into China.

POLLINATION   The flower structure of *Coelogyne* suggests bee pollina-tion; Tom Reeves (pers. comm.) reports that *Coelogyne fragrans* is pollinated by wasps.

CHROMOSOME NUMBER   40.

SPECIES   About 440.

GENERA   16: *Acoridium, Basigyne, Bulleyia, Coelogyne, Dendrochilum, Dickasonia, Gynoglottis, Ischnogyne, Nabaluia, Neogyne, Otochilus,*

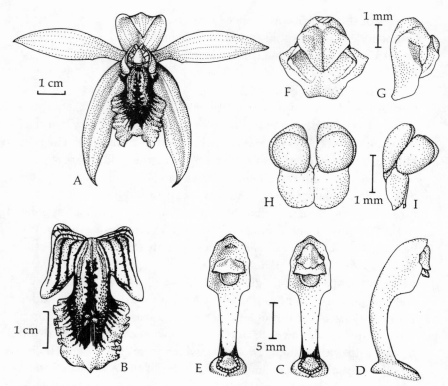

Figure 9.5 *Coelogyne pandurata* (Epidendroideae: Coelogyneae). (*A*) Flower, front view. (*B*) Lip, flattened. (*C*) Column, ventral view. (*D*) Column, side view. (*E*) Column, after removal of anther. (*F*) Anther, ventral view. (*G*) Anther, side view. (*H*) Pollinia, top view. (*I*) Pollinia, side view.

Panisea, Pholidota, Pleione, Pseudacoridium, Sigmatogyne.

DISCUSSION   This is an unusually clear-cut group, easily recognized by either habit or flower structure. In many cases the inflorescence appears to be lateral, but as far as I know it is always terminal on a pseudobulb that develops later.

REFERENCES   Butzin, 1974 (keys to cultivated taxa); Pfitzer and Kränzlin, 1907 (revision); Seidenfaden, 1975c (*Coelogyne* in Thailand).

Subtribe ADRORHIZINAE Schlechter

DESCRIPTION   Habit: small epiphytes without pseudobulbs but with very fleshy roots; vegetative stems very short. Leaves: duplicate, articulate. Inflorescence: terminal(?), of one or few spiral flowers. Flowers: small, resupinate; column slender, anther terminal, incumbent; four pollinia, superposed; stigma entire.

DISTRIBUTION   Tropical Asia (southern India and Ceylon).

POLLINATION   Not known.

CHROMOSOME NUMBER   Not known.

SPECIES   3.

GENERA   2: *Adrorhizon, Sirhookera.*

DISCUSSION   These little plants have a distinctive habit, but the super-posed pollinia are very similar to those of *Coelogyne.*

Tribe MALAXIDEAE Lindley

DESCRIPTION   Habit: terrestrial or epiphytic, with pseudobulbs or corms of one or several internodes, or stems slender. Leaves: distichous, con-volute or duplicate, plicate, conduplicate or laterally flattened and fleshy, articulate or not. Inflorescence: terminal, simple, of few to many flowers, spiral or distichous. Flowers: small to medium-small, resupinate or not; column short or elongate; anther terminal or subdorsal, incumbent or more or less erect, two-celled; four pollinia, hard, oblong or usually somewhat clavate; stigma entire, often with two small viscidia. (See fig. 9.6.)

DISTRIBUTION   World-wide (except New Zealand).

POLLINATION   There are almost no observations; fly pollination and pollination by a bug have each been reported once in *Liparis*, but this is scarcely enough to give us much idea of the floral ecology of this group.

CHROMOSOME NUMBERS   20, 26, 28, 30, 38, 42, 44.

SPECIES   About 890.

GENERA   6: *Hippeophyllum, Liparis, Malaxis, Oberonia, Orestias, Risleya.*

DISCUSSION   This tribe is characterized by "naked" pollinia. In a family in which the whole classification is traditionally based on the details of the pollinia and associated features, this means that the tribe is taxo-nomically poverty stricken. Even the line between *Malaxis* and *Liparis* is unclear. There are probably groups within each "genus" that are much more distantly related to each other than are the genera that we recog-nize in other tribes, but these cannot be delineated on the basis of our present knowledge.

There is, of course, no very satisfactory distinction between the Malaxideae and the Dendrobiinae, but they seem to be separate groups, and probably only very distantly related. The Malaxideae often have plicate leaves, the flowers have little or no column foot, and the pollinia are usually pointed at the rostellar end, though there are no distinct caudicles. There is a general resemblance between this group and the Listerinae, but it is probably due to convergence rather than phyletic relationship.

RELATIONSHIPS   This tribe is thought to be more closely related to the

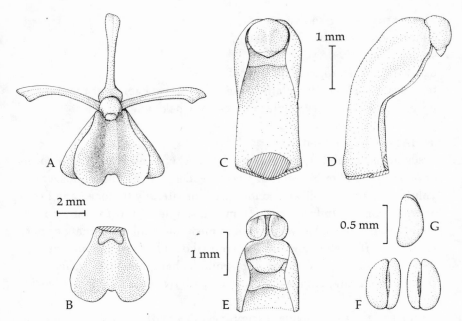

Figure 9.6 *Liparis nervosa* (Epidendroideae: Malaxideae). *(A)* Flower, front view. *(B)* Lip, flattened. *(C)* Column, ventral view. *(D)* Column, side view. *(E)* Apex of column, with anther tipped back. *(F)* Pollinia. *(G)* Pollinia, side view.

Arethuseae than to the Epidendreae, as indicated by Lavarack (1971), and the frequent presence of plicate leaves seems to fit this idea.

PHYLETIC TRENDS   I would guess that the elongate column of *Liparis* is more primitive than the short column and nearly erect anther of *Malaxis* or *Oberonia*, and the pleated leaves also would be a primitive feature. Thus, the many epiphytic species of Asia are more advanced than some of the terrestrials.

REFERENCES   Seidenfaden, 1968 (*Oberonia*), 1976 (*Liparis* in Thailand), 1978b (*Oberonia* and *Malaxis* in Thailand).

Tribe CRYPTARRHENEAE Dressler, 1980b

DESCRIPTION   Habit: epiphytic, with pseudobulbs or short, unthickened stems. Leaves: distichous, duplicate, articulate. Inflorescence: lateral, simple, of many spiral flowers. Flowers: small, resupinate; lip clawed, with a thick callus on claw, four-lobed; column short, with a hooded clinandrium over the anther; anther terminal, incumbent but with an erect beak, two-celled; four pollinia, superposed, with two hyaline cylindrical caudicles that may be basally joined, and a distinct viscidium; stigma entire. (See fig. 9.7.)

DISTRIBUTION Tropical America.

POLLINATION Not known.

CHROMOSOME NUMBER Not known.

SPECIES 3 or 4.

GENUS *Cryptarrhena.*

DISCUSSION This is one of the most isolated orchid genera. It has been listed in Ornithocephalinae, with which it agrees only in having four pollinia. It is not even altogether clear whether it should be placed in the Epidendroideae or the Vandoideae. The main part of the anther appears incumbent, though the beak is rather erect, and there is no stipe; so the Epidendroideae seems a better choice. It seems about as much like the Coelogyneae as any other group but no close relationships can be indicated.

REFERENCE Dressler, 1980b (relationships).

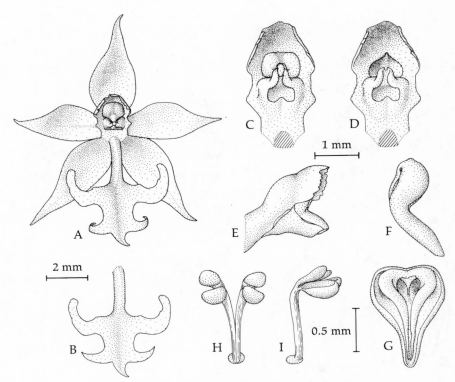

Figure 9.7 *Cryptarrhena guatemalensis* (Epidendroideae: Cryptarrheneae). (*A*) Flower, front view. (*B*) Lip, flattened. (*C*) Column, ventral view. (*D*) Column, with anther and pollinarium removed. (*E*) Column, side view. (*F*) Anther, side view. (*G*) Anther, ventral view. (*H*) Pollinarium, top view. (*I*) Pollinarium, side view.

Tribe CALYPSOEAE Dressler, 1979b

DESCRIPTION   Habit: terrestrial or saprophytic. Leaves: one (or absent), convolute, plicate, petiolate, nonarticulate. Inflorescence: lateral(?), simple, of few spiral flowers. Flowers: medium, resupinate, lip deeply saccate, column broad, anther ventral, incumbent, with poorly developed partitions; four pollinia, in two pairs, superposed, on a prominent viscidium; stigma entire. (See fig. 9.8.)

DISTRIBUTION   North temperate (North America and eastern Asia).

POLLINATION   The genus *Calypso* is usually pollinated by bumblebees. In some populations the cluster of hairs on the lip is yellow and apparently serves to simulate pollen-bearing anthers and attract naive bees.

CHROMOSOME NUMBERS   28, 32.

SPECIES   3.

GENERA   2: *Calypso, Yoania.*

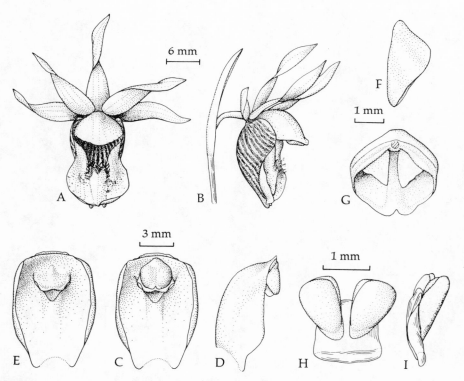

Figure 9.8   *Calypso bulbosa* (Epidendroideae: Calypsoeae). (*A*) Flower, front view. (*B*) Flower, side view. (*C*) Column, ventral view. (*D*) Column, side view. (*E*) Column, with anther and pollinarium removed. (*F*) Anther, side view. (*G*) Anther, ventral view. (*H*) Pollinarium, top view (*I*) Pollinarium, side view. All except *H* and *I* drawn from material preserved in liquid.

DISCUSSION   These two genera have not been associated in other classifications, but their flower structure is very similar. Brieger (in Schlechter,
1970) places the Calypsoinae in the Epidendroideae because they supposedly lack a stipe. Actually, *Calypso* has a well-developed stipe, but
the incumbent anther and the emergent stigma indicate that *Calypso* is
correctly placed in the Epidendroideae. Its relationship seems to be with
the Arethuseae and the Coelogyneae. It is quite possible that other
genera listed in the Corallorhizinae, such as *Dactylostalix*, will prove to
belong here, though they do not show the broad, flattened column of
Calypso and *Yoania*.

RELATIONSHIPS   Probably with the Arethuseae and Coelogyneae.

Tribe EPIDENDREAE Humboldt, Bonpland, and Kunth

DISCUSSION   In floral features, some of the Epidendreae are very like
some of the Arethuseae, but they are more distinctive in vegetative
features. Many Epidendreae have slender stems with distichous, conduplicate leaves. Very few have clearly pleated leaves. The inflorescence
is usually terminal or upper axillary. The primitive members have eight
pollinia which may be either laterally flattened or clavate. Most subtribes include genera with the primitive number of pollinia, and there is
usually a reduction series to six, four, or two pollinia. In many cases we
find the development of viscidia, especially associated with clavate pollinia (see fig. 9.9). The Dendrobiinae and Bulbophyllinae resemble the
Malaxideae in the possession of naked pollinia, but they appear to be
closely allied to the Eriinae. Brieger (in Schlechter, 1975) separates some
subtribes as the tribe Podochileae and characterizes the group by the
presence of a viscidium. I am unable to find any consistent difference
between the Podochileae and the Epidendreae (in the sense of Brieger).
Brieger seems to consider the "Podochileae" a primitive group because
of its superficial resemblance to the Spiranthoideae. Indeed, he suggests
that the Arethuseae and the "Podochileae" are both primitive groups
with respect to the Epidendroideae, but that they have evolved independently of each other. I cannot visualize the evolution of a natural
subfamily with two unrelated primitive groups.

RELATIONSHIPS   The Epidendreae seem clearly related to the Arethuseae,
and surely have a common ancestry with that group.

PHYLETIC TRENDS   There are several interesting trends in this tribe,
mostly occurring in a parallel manner in several subtribes. The pollinia
are primitively eight, but may be reduced in number, whether laterally
flattened or clavate, and, in some of the more derived groups, prominent
caudicles simulate the stipes of the vandoid orchids (fig. 9.9). In the

Dendrobiinae and Bulbophyllinae, on the other hand, the pollinia are usually quite naked, but viscidia occur in some species of *Bulbophyllum*, and stipes are known in *Monomeria* and *Sunipia*. Most groups show the development of pseudobulbs, generally as slender pseudobulbs of several internodes, but some consist of a single internode. In the Pleurothallidinae no pseudobulbs are developed, the leaves themselves being fleshy in the more xeric representatives.

Subtribe ERIINAE Bentham

DESCRIPTION   Habit: epiphytic, rarely terrestrial; stem slender or forming pseudobulbs, usually of several internodes. Leaves: distichous (secondarily spiral in some dwarf species), duplicate (rarely convolute and plicate), articulate. Inflorescence: terminal, or usually lateral, often upper axillary, simple, of several to many spiral flowers. Flowers: resupinate or not, small to medium-small; sepals free or united, sometimes with a prominent spur; column usually with a prominent column foot; anther terminal, incumbent, eight-celled; eight pollinia, laterally flattened or clavate, with intralocular caudicles; stigma entire or somewhat two-lobed, often with a distinct viscidium.

DISTRIBUTION   Tropical Asia and Australasia, *Stolzia* in Africa.

POLLINATION   Not known; *Cryptochilus, Mediocalcar,* and *Porpax* show features suggesting bird pollination. Many species of *Eria* have pseudo-pollen on the lip and could well be pollinated by small bees.

CHROMOSOME NUMBERS   36, 38, 40, 44.

SPECIES   About 500.

GENERA   8 in two tentative alliances: (1) *Cryptochilus, Eria, Mediocalcar, Porpax, Stolzia.* (2) *Ceratostylus, Epiblastus, Sarcostoma.*

DISCUSSION   This group is very similar to the New World Laeliinae, and the main distinction seems to be geographic. However, the Eriinae usually have lateral inflorescences, and they seem to be a real phyletic group, in spite of the general resemblance. Brieger (in Schlechter, 1975) places *Cryptochilus* and *Mediocalcar* together with *Glomera*, because they possess viscidia, but viscidia are found in many *Eria* species (but not in *Glomera*), and the eight pollinia of these genera would be discordant in the Glomerinae, but not at all so in the Eriinae. *Ceratostylus, Epiblastus,* and *Sarcostoma* have been included, traditionally, with *Glomera*, but their habit is different, and the column structure and pollinia discordant. It is possible that they should be placed in a separate subtribe. The African *Stolzia* was compared to *Polystachya* by Schlechter (1926), but the details of the column and pollinia coincide completely with the Eriinae. The habit, however, is somewhat distinctive.

REFERENCES   Cribb, 1978 (revision of *Stolzia*); Kränzlin, 1911 (taxonomic revision of *Eria*, inadequate).

Subtribe PODOCHILINAE Bentham and Hooker

DESCRIPTION   Habit: epiphytic or terrestrial, with slender, reedlike stems or sometimes with pseudobulbs. Leaves: distichous, duplicate, sometimes laterally flattened, articulate or not. Inflorescence: terminal or lateral, of few to many spiral flowers. Flowers: small, resupinate or pendant, lip often basally saccate or spurred; column short, often with a prominent foot; anther terminal and incumbent or dorsal and erect; eight, six, or four pollinia, clavate, often with one or two prominent caudicles or with abortive pollinia at base; stigma entire, with one or two viscidia.

DISTRIBUTION   Tropical Asia and Australasia with *Agrostophyllum* represented in the Seychelles (Africa).

POLLINATION   Not known.

CHROMOSOME NUMBER   Not known.

SPECIES   About 230.

GENERA   7: *Agrostophyllum, Appendicula, Chilopogon, Chitonochilus, Cyphochilus, Poaephyllum, Podochilus.*

DISCUSSION   The Glomerinae and Podochilinae are traditionally separated by the position of the anther, but I find this inconsistent in both groups, however delineated. The nature of the pollinia, on the other hand, seems to separate two distinctive groups. As here delineated, the Podochilinae are characterized by clavate pollinia and a transverse plate near the base of the lip. Some species of *Appendicula* and *Podochilus* (but by no means all) have very stipelike structures, but these are abortive pollinia, rather than stipes or caudicles (fig. 9.9).

Subtribe THELASIINAE Schlechter

DESCRIPTION   Habit: epiphytic, with or without pseudobulbs, stems, if slender, either short or elongate. Leaves: distichous, duplicate, often laterally flattened and fleshy, articulate or not. Inflorescence: lateral, of few to many spiral flowers. Flowers: tiny, resupinate or not; column short, with or without a foot; anther erect, dorsal; eight pollinia, ovoid, with a common caudicle, this sometimes very long; stigma entire, with a viscidium.

DISTRIBUTION   Tropical Asia and Australasia.

POLLINATION   Not known.

CHROMOSOME NUMBER   Not known.

SPECIES   About 270.

GENERA   7: *Chitonanthera, Octarrhena, Oxyanthera, Phreatia, Rhynchophreatia, Ridleyella*(?), *Thelasis.*

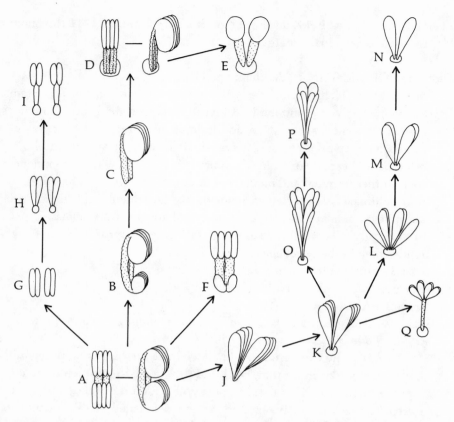

Figure 9.9 Patterns of pollinia evolution in the Epidendreae. *(A–I)* Laterally flat-tened pollinia. *(J–Q)* Clavate pollinia. *(D,E,H,I, K–Q)* Shown with viscidia. *I* has stipes, as in *Sunipia*; *O* and *P* have abortive pollinia, as in the Podochilinae; and *Q* has long caudicles, as in the Thelasiinae.

DISCUSSION These genera are characterized by ovoid pollinia attached to a common caudicle, and rather minute flowers. In some cases, the very long caudicle has been interpreted as a stipe. *Ridleyella* was compared to *Bulbophyllum* and assigned a separate subtribe, but the eight pollinia are quite pointed, and I believe that it belongs with this group.

REFERENCE Kränzlin, 1911 (taxonomic revision, inadequate).

Subtribe GLOMERINAE Schlechter

DESCRIPTION Habit: epiphytic or terrestrial, with slender, reedlike stems or sometimes with pseudobulbs. Leaves: distichous, duplicate, articulate. Inflorescence: terminal, of few to many spiral flowers, often dense. Flowers: small to medium, resupinate or not; lip often basally saccate or spurred; column short, often with a prominent foot; anther terminal, incumbent or more or less dorsal and erect; four pollinia, laterally flattened, with small caudicles and with or without viscidium; stigma entire.

DISTRIBUTION   Tropical Asia and Australasia.

POLLINATION   Not known; in some cases the flower structure is very suggestive of moth pollination.

CHROMOSOME NUMBER   Not known.

SPECIES   About 130.

GENERA   6: *Aglossorhyncha, Earina, Glomera, Glossorhyncha, Ischnocentrum, Sepalosiphon.*

DISCUSSION   This group is characterized by four laterally flattened pollinia with inconspicuous caudicles. I find no viscidium in *Glomera*, but some other genera do have an ill-defined viscidium (semiliquid). This group has been confused with the Podochilinae, but may not be very closely allied with that subtribe. *Earina* has pseudobulbs in some species, and an unusual, branched inflorescence, the branches being condensed and producing flowers over a long period.

Subtribe LAELIINAE Bentham

DESCRIPTION   Habit: epiphytic or terrestrial, stems slender or forming pseudobulbs, these usually of several internodes. Leaves: distichous, or terminal on pseudobulb, duplicate, usually articulate. Inflorescence: terminal or rarely lateral, simple or branched, of one to many flowers, spiral or distichous. Flowers: tiny to large, resupinate or not; flowers may have a cuniculus type of nectary; column short or elongate, often winged, may have a column foot; anther terminal and incumbent or erect; pollinia laterally flattened or ovoid, eight, six, four, or two, with prominent caudicles; stigma entire, sometimes with a viscidium. (See fig. 9.10.)

DISTRIBUTION   Tropical America.

POLLINATION   For this subtribe we have better, but still very incomplete, knowledge indicating a considerable adaptive radiation. The flowers of *Cattleya, Laelia* (exc. section *Parviflorae*), *Schomburgkia*, and most *Encyclia* species suggest bee pollination, and that is what we have observed, with wasp pollination in some *Encyclia* species. Braga (1978) reports that *Caularthron* is pollinated by carpenter bees (*Xylocopa*). *Brassavola* and *Rhyncholaelia* both are usually pollinated by sphingid moths. *Isochilus, Laelia* section *Parviflorae, Alamania, Neocogniauxia, Sophronitis,* and *Hexisea* all have the aspect of hummingbird flowers. *Epidendrum* is basically Lepidopteran pollinated, but we find a few

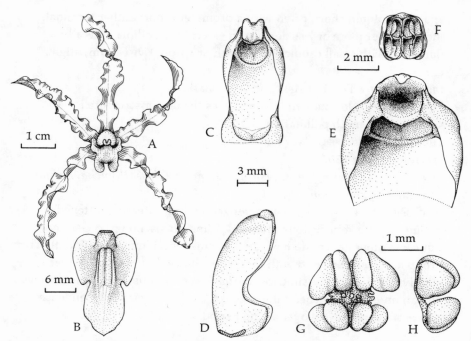

Figure 9.10 *Schomburgkia lueddemannii* (Epidendroideae: Epidendreae). *(A)* Flower, front view. *(B)* Lip, flattened. *(C)* Column, ventral view. *(D)* Column, side view. *(E)* Apex of column, with anther and pollinarium removed. *(F)* Anther. *(G)* Pollinarium, top view. *(H)* Pollinarium, side view.

species adapted to fly pollination and several that are adapted to hummingbird pollination. N. H. Williams (pers. comm.) has found stingless bees *(Trigona)* pollinating the flowers of *Scaphyglottis*.

CHROMOSOME NUMBERS   24, 40, 56. Forty is the usual chromosome number for this subtribe, but the members of the *Epidendrum ibaguense-secundum* complex have shown different numbers in some species. More counts of this complex are needed.

SPECIES   About 830.

GENERA   43 in six tentative alliances: (1) *Helleriella, Hexisea, Isochilus, Nidema, Platyglottis, Ponera, Reichenbachanthus, Scaphyglottis.* (2) *Artorima, Basiphyllaea, Brassavola, Broughtonia, Cattleya, Constantia, Encyclia, Hagsatera, Isabelia, Laelia, Pseudolaelia, Quisqueya, Rhyncholaelia, Schomburgkia, Sophronitis, Tetramicra.* (3) *Barkeria, Caularthron, Orleanesia.* (4) *Dimerandra, Diothonaea, Epidanthus, Epidendrum, Jacquiniella, Neowilliamsia, Oerstedella.* (5) *Alamania, Domingoa, Homalopetalum, Leptotes, Loefgrenianthus, Nageliella, Pinelia.* (6) *Dilomilis, Neocogniauxia.*

INTERGENERIC HYBRIDS   There are a number of natural hybrids known in the Laeliinae, and some of them are intergeneric by current standards (fig. 9.11). Natural hybrids between *Cattleya* and *Brassavola* are not common, but three are recorded. Hybrids between *Cattleya* and *Laelia*, on the other hand, are surprisingly frequent, nineteen different × *Laeliocattleya* hybrids being recorded in nature. Quite apart from the high degree of interfertility, there are good grounds for reclassifying some or all sections of *Laelia* into the genus *Cattleya*. There are a few natural hybrids known between *Broughtonia* and *Encyclia* in the West Indies, and one hybrid between *Hexisea* and *Scaphyglottis* has been found in Panama.

The number of artificial intergeneric hybrids is far greater. Most of the genera in the *Cattleya* alliance are easily crossed, and in many cases

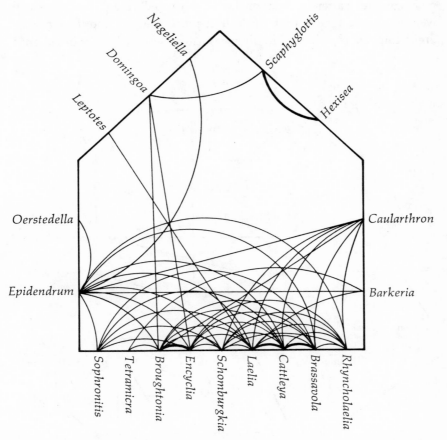

Figure 9.11   Intergeneric hybrids known in the subtribe Laeliinae. The heavy lines indicate naturally occurring hybrids.

the hybrids are quite fertile. Both *Caularthron* and *Epidendrum* have been crossed with several other genera, the *Epidendrum* hybrids, at least, being largely sterile. *Domingoa, Leptotes,* and *Nageliella* have each been crossed with other genera, and one cross between *Scaphyglottis (Hexadesmia)* and *Domingoa* is registered. It is probable that most intergeneric combinations in this subtribe will yield viable offspring, though some of the combinations may be sterile.

DISCUSSION Pfitzer (1887) and Schlechter (1926) both divided this group into two subtribes, the *Scaphyglottis* group (Ponerinae), with a prominent column foot, and the *Cattleya* group (Laeliinae) without a column foot, but this distinction has not been so simple, especially with the genera that have, or seem to have, a slight column foot. Both Schlechter and Brieger have interpreted the saccate nectary of *Jacquiniella,* for example, as being a column foot. Baker's anatomical study suggests a more complex pattern of relationships, with two different complexes with well-developed column foot, the *Scaphyglottis* complex and the *Domingoa* complex (fig. 9.12). The alliances listed here are largely based on

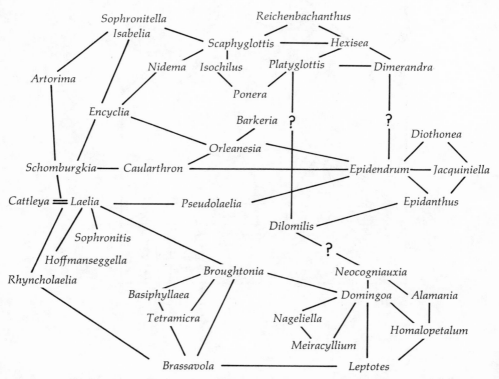

Figure 9.12 Suggested anatomical relationships in the Laeliinae. (After Baker, 1972.)

Baker's study, with most doubtful genera placed in the already ill-defined *Cattleya* alliance. The classification of this subtribe in the new edition of Schlechter is highly artificial.

Dilomilis, with eight pollinia and a slender reedlike stem, may represent the most primitive living genus of this subtribe. I had earlier excluded *Isochilus* from this subtribe because of its resemblance to *Elleanthus*, but Baker's study indicates that it belongs to the Laeliinae, the resemblance to *Elleanthus* being convergence due to hummingbird pollination.

REFERENCES Baker, 1972 (anatomy); Fowlie, 1977 (bifoliate *Cattleya* species).

Subtribe MEIRACYLLIINAE Dressler

DESCRIPTION Habit: epiphytic, stems short, slightly thickened. Leaves: one per growth, duplicate, fleshy, articulate. Inflorescence: terminal, simple, of few spiral flowers. Flowers: small, resupinate, lip basally saccate; column elongate, anther erect, dorsal, eight-celled; eight pollinia, clavate, with a distinct viscidium; stigma entire.

DISTRIBUTION Tropical America (Mexico and Central America).

POLLINATION Not known; the perfume of methyl cinnamate suggests pollination by euglossine males.

CHROMOSOME NUMBER Not known.

SPECIES 2.

GENUS *Meiracyllium*.

DISCUSSION This genus has been placed in the Laeliinae, where its beaklike rostellum and long, clavate pollinia seem out of place, though Baker (1972) found the anatomy to be concordant with the Laeliinae.

REFERENCE Dressler, 1960 (relationships).

Subtribe PLEUROTHALLIDINAE Lindley

DESCRIPTION Habit: epiphytic or terrestrial, without pseudobulbs, stems unifoliate. Leaves: duplicate, often fleshy, articulate. Inflorescence: terminal, or rarely lateral; simple or fascicled, distichous. Flowers: resupinate or not, with a joint between ovary and pedicel; column short or elongate, often with a distinct foot, anther apical and incumbent or dorsal and erect; pollinia clavate, eight, six, four, or two, often with a tiny viscidium (when anther is erect); stigma entire or two-lobed; capsule may have two unequal valves.

DISTRIBUTION Tropical America.

POLLINATION This is basically a fly-pollinated group. Most of our observations are of small *Drosophila*-like flies, and most genera have flowers which show an appropriate syndrome. Some of the *Masdevallia* species are pollinated by carrion flies, and a few others by humming-birds. Vogel (1978) suggests that the remarkably fungus-like lip of *Dracula* represents mimicry and that the flowers are pollinated by fungus flies.

CHROMOSOME NUMBERS   32, 42, 44, 64; few counts are known.

SPECIES   About 3,800.

GENERA   26: *Acostaea, Andreettaea, Barbosella, Brachionidium, Chamelophyton, Cryptophoranthus, Dracula, Dresslerella, Dryadella, Lepanthes, Lepanthopsis, Masdevallia, Octomeria, Phloeophila, Physosiphon, Physothallis, Platystele, Pleurothallis, Porroglossum, Restrepia, Restrepiella, Restrepiopsis, Salpistele, Scaphosepalum, Stelis, Triaristella.*

INTERGENERIC HYBRIDS   Two artificial hybrids between *Dracula* and *Masdevallia* are registered. I doubt that anyone has attempted crossing the other, small-flowered members of this subtribe.

DISCUSSION   This group is easily recognized by its distinctive habit and by the joint between the ovary and the pedicel. This is one group in which differences in number of pollinia seem to correspond very well with generic boundaries. There are yet some artificialities in the classification of this subtribe. Hopefully, these will be clarified by studies of anatomy and other features in the near future.

REFERENCES   Garay, 1974c (general); Kränzlin, 1925 (revision of *Masdevallia*); Misas and Arango, 1974 (general); Vogel, 1978 (pollination of *Dracula*); see also the first few volumes of Selbyana for a series of taxonomic articles by Luer.

Subtribe DENDROBIINAE Lindley

DESCRIPTION   Habit: epiphytic or occasionally terrestrial, stems slender or forming pseudobulbs, these usually of several internodes, sometimes of a single internode. Leaves: distichous, duplicate, sometimes laterally flattened, articulate. Inflorescence: lateral or terminal(?), usually upper axillary, simple or branched, of few to many spiral flowers. Flowers: small to large, resupinate; lip often jointed basally, flowers often with a spur formed by the column foot or by the lip and column foot; column with a prominent foot; anther terminal and incumbent, two-celled; four or two pollinia, naked, in two pairs, without caudicles or viscidia.

DISTRIBUTION   Tropical Asia and Australasia.

POLLINATION   Most of the records for *Dendrobium* indicate bee pollination, though syrphid flies and thynnid wasps are also mentioned. One case of bird pollination is recorded, but a number of the species of higher elevations show the bird-pollination syndrome. A number of species show ephemeral flowers and flower gregariously.

CHROMOSOME NUMBERS   38, 40.

SPECIES   About 1,650.

GENERA   6: *Cadetia, Dendrobium, Diplocaulobium, Ephemerantha, Epigeneium, Pseuderia.*

DISCUSSION   This subtribe is characterized by having naked pollinia, without any trace of caudicles. In a family in which the whole classification is traditionally based on the details of the pollinia and associated features, the taxonomy of the group is almost inevitably rather poor. In this group we find four or sometimes two pollinia that are more or less ellipsoid, only rarely somewhat attenuate at the rostellar end. Naked pollinia seem to place definite functional limits on the kinds of floral diversity, and even more severe constraints on a taxonomy that is based primarily on the pollinia. Holttum (1952b) has indicated that many of the intersterile sections of *Dendrobium* are biologically far more distinctive than the rapidly multiplying genera of the Sarcanthinae. Unfortunately, even the sectional classification needs revision. As presently treated, *Dendrobium* is one of the largest orchid genera, with an estimated 1,400 species.

While *Dendrobium* lacks a distinct viscidium, some of the seemingly bird-pollinated species, such as *D. secundum* and *D. sophronites,* show what must be the penultimate step in the evolution of a viscidium. In these species, a not-too-sharply-defined piece of the rostellum comes away as a unit (Reichenbach, 1884). The pollinia are not attached to the rostellum but are very efficiently caught by the sticky "semiviscidium" when the anther tips back and the pollinia touch this freshly removed piece of the rostellum.

The main difference between this subtribe and most other Epidendreae is the same one that sets the Malaxideae apart as a tribe. Here, however, the Dendrobiinae resemble the other Epidendreae, and especially the Eriinae, in chromosome number, habit, seed structure, and general flower structure. Many authors have placed *Dendrobium* and *Eria* in the same subtribe. The structure of the pollinia in these two genera is quite different, though they are surely closely related.

REFERENCES   Holttum, 1952b (sections); Jones and Gray, 1976b, 1977 (pollination); Kränzlin, 1910 (revision, inadequate).

Subtribe BULBOPHYLLINAE Schlechter

DESCRIPTION   Habit: epiphytic, with pseudobulbs of a single internode, these often widely separated on rhizome, and sometimes reduced in size. Leaves: duplicate, often fleshy, articulated, sometimes reduced to scales. Inflorescence: lateral, simple, spiral or distichous, of one to many flowers. Flowers: small to large, resupinate; lip often hinged at base;

column with a prominent foot; anther terminal, incumbent, two-celled; four or two pollinia, naked, occasionally with a viscidium or viscidia or even a stipe; stigma entire.

DISTRIBUTION   Pantropical, especially in the Old World tropics.

POLLINATION   Most *Bulbophyllum* species seem to conform to the fly-pollination syndrome, and our records for Asiatic species all indicate fly pollination. The perfume of some species leaves no doubt of carrion-fly pollination; some species, however, produce nectar (Jones and Gray, 1976a). Johansson's observations in West Africa (1974), surprisingly, indicate wasps, stingless bees, and ctenuchid wasps as visitors of three *Bulbophyllum* species. If these are, indeed, pollinators of West African *Bulbophyllum*, it shows us, once again, that we should not place too much trust in the syndromes.

CHROMOSOME NUMBERS   36, 38, 40.

SPECIES   About 1,020.

GENERA   7: *Bulbophyllum, Chaseëlla, Drymoda, Monomeria, Pedilochilus, Saccoglossum, Trias.*

DISCUSSION   This group, too, is characterized by naked pollinia, and there is little agreement on which groups merit generic status and which do not. Here, however, something has been added. Some species of *Bulbophyllum* have definite viscidia, though they seem to have no caudicles. The genera *Drymoda, Monomeria,* and *Sunipia* have been placed in a separate subtribe with the African *Genyorchis,* all supposedly having stipe or stipes. Unfortunately, *Genyorchis* does not belong to this group, as its column and pollinary structure place it in the Cymbidieae. *Drymoda* has a well-developed viscidium as do some species of *Bulbophyllum,* but no stipe. Its vegetative features are reminiscent of *Porpax* (Eriinae) and it is not closely allied to either *Monomeria* or *Sunipia. Monomeria* has a stipe and is otherwise similar to *Bulbophyllum. Sunipia* is more distinctive, and is treated below in a separate subtribe.

REFERENCES   Jones and Gray, 1976a (pollination); Seidenfaden, 1973 (section *Cirrhopetalum*).

Subtribe SUNIPIINAE Dressler, 1979b

DESCRIPTION   Habit: epiphytic, with pseudobulbs of a single internode, these usually widely separated on rhizome. Leaves: duplicate, fleshy, articulate. Inflorescence: lateral, simple, spiral or distichous, of one to many flowers. Flowers: small to medium, resupinate; lip simple; column with a short foot; anther terminal, united with the clinandrium, opening away from the clinandrium, two-celled; four pollinia, somewhat laterally flattened, each pair on a separate stipe with a viscidium; stigma entire.

DISTRIBUTION Northern tropical Asia.

POLLINATION Not known.

CHROMOSOME NUMBER Not known.

SPECIES About 25.

GENUS *Sunipia.*

DISCUSSION This genus shows much the habit of *Bulbophyllum* and is presumably allied to that genus, but the floral details are very distinctive. The anther and the paired stipes are both unique in this subfamily.

REFERENCE Seidenfaden, 1969 (taxonomic revision).

The Vandoid Orchids

10

Whether treated as a tribe, a subfamily, or an evolutionary level, the vandoid group has been recognized in some way by most authors, though there is little agreement on exactly where to draw the line between epidendroid and vandoid orchids. There are a number of features which are usually associated with the vandoid group: lateral inflorescence, anther with reduced partitions, superposed pollinia, viscidium and stipe, but these features are not always consistent in their correlations. We find many epidendroid orchids with lateral inflorescences, and several undoubted vandoids with terminal inflorescences. The anthers of *Sobralia* and *Coelogyne* are strikingly vandoid in character, as are the pollinia of *Coelogyne* (but without either viscidium or stipe). A viscidium is found in many orchid groups, and it has undoubtedly evolved independently in a number of them. A stipe is found in *Prasophyllum*, *Sunipia*, and *Monomeria*, though none of them can possibly be considered as vandoid. Similarly, I interpret *Calypso* as a member of the Epidendroideae which has independently evolved a stipe and, indeed, quite vandoid pollinaria. While most vandoid orchids have a stipe or stipes, it is very small or absent in some species of *Maxillaria* and in some members of the Corallorhizinae and Cyrtopodiinae. Brieger (in Schlechter, 1970) places the Corallorhizinae in the Epidendroideae because they supposedly lack a stipe (but see discussion under Calypsoeae).

I believe that the only feature which gives a satisfactory division between the Epidendroideae and the Vandoideae is the development of the anther. While Hirmer gave details of this feature in 1920, he himself misinterpreted his observations, as have the rest of us since them. After all, a caplike anther sitting on the apex of the column looks much the same in *Cattleya* or in *Cymbidium*; it is difficult to convince oneself that there is a fundamental difference. Yet the anther of *Cattleya* bends down as it develops, and that of *Cymbidium* does not. Hirmer interpreted the anther of *Cymbidium* as being already bent in its earliest stages. After comparing the development of many vandoid and other orchids, I believe, rather, that the anther of *Cymbidium* and other vandoid orchids is erect, very short, and opens basally, rather than ventrally. This is much more obvious in the flower of *Neobenthamia* than in most other vandoid orchids. By using this feature, in addition to all the others, we can make a clear distinction between the Epidendroideae and the Vandoideae, and once we recognize the fundamental difference, it seems only reasonable to treat them as distinct subfamilies.

Subfamily
VANDOIDEAE Endlicher

DISCUSSION   By tradition, the vandoids are considered the most highly evolved orchids. The anther is more modified than in most other groups, the pollinia are quite hard and are associated with rostellar and columnar tissues to form a rather complex "pollinarium," which is sufficiently complex that many genera of vandoid orchids may be recognized by their pollinaria alone. At the same time, the flower structure is basically very similar in all vandoid groups, and dividing the subfamily into tribes is a bit arbitrary. I divide the subfamily into four tribes, though there are a few subtribes which could easily be elevated to tribal status, and it is difficult (without more knowledge of anatomy, cytology, and biochemistry) to show convincingly that one system is superior to another.

In this subfamily the number of pollinia seems to be a bit more important than in others, but the pollinaria must be observed very carefully. When there are only two pollinia, each pollinium is usually deeply cleft on one or two sides, and solid on one side only, so that it looks very much like a pair of superposed pollinia. I believe that four pollinia is the primitive condition here, and that reduction to two has occurred several times. Indeed, this is a very critical question, and I will try to indicate those groups that seem doubtful to me.

RELATIONSHIPS   The Vandoideae show some resemblances to the Epidendroideae and the Orchidoideae, but there is a large gap between the Vandoideae and its nearest relatives, whatever these may be.

Tribe POLYSTACHYEAE Pfitzer

DESCRIPTION   Habit: epiphytic or terrestrial; stems slender or forming pseudobulbs, these usually of several internodes, sometimes of a single internode. Leaves: distichous, duplicate, articulate. Inflorescence: terminal or lateral, simple or branched, of several to many spiral flowers. Flowers: small to medium-small, resupinate or not, lip commonly with mealy hairs (pseudopollen) on upper surface; column with a prominent foot, short or somewhat elongate; anther terminal, operculate, with reduced partitions; four pollinia, superposed, with a small but definite stipe and a viscidium; stigma entire. (See fig. 10.1.)

DISTRIBUTION   Pantropical, but primarily African, with a secondary center in tropical America.

POLLINATION   The genus *Polystachya* is known to be pollinated by small

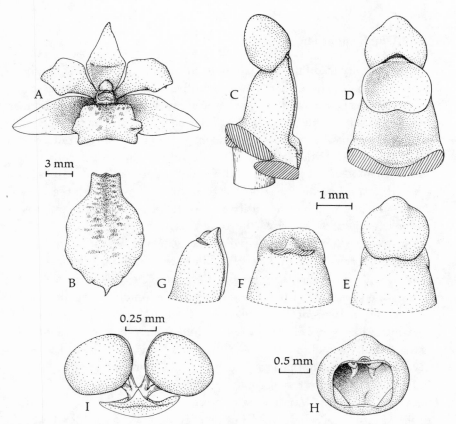

Figure 10.1   *Neobenthamia gracilis* (Vandoideae: Polystachyeae). *(A)* Flower. *(B)* Lip, flattened. *(C)* Column, side view. *(D)* Column, ventral view. *(E)* Apex of column with anther in place, dorsal view. *(F)* Apex of column with anther and pollinarium removed, dorsal view. *(G)* Apex of column with anther and pollinarium removed, side view. *(H)* Anther, basal (clinandrial) view. *(I)* Pollinarium.

bees that gather the pseudopollen. This is probably the pollination system of most species of this subtribe.

CHROMOSOME NUMBER   40.

SPECIES   About 220.

GENERA   4: *Hederorkis, Imerinaea, Neobenthamia, Polystachya*.

DISCUSSION   The taxonomic vicissitudes of this group show nicely the problems involved in the taxonomy of the vandoid orchids. Pfitzer (1887) and Schlechter (1926) grouped with *Polystachya* most genera of the Cyrtopodiinae with terminal inflorescence, while Dressler and Dodson (1960) and Garay (1960), at the other extreme, excluded *Polystachya* from the vandoid groups altogether. Though the viscidium and the stipe of the Polystachyeae are small, they are quite distinct, and these and all other features indicate that these are undoubtedly vandoid by any criterion. The elongate, clearly erect anther and the curious calyculus of *Neobenthamia* suggest that it is one of the more primitive genera of this subfamily.

RELATIONSHIPS   The Polystachyeae may be more closely related to the Vandeae than to the primarily cormous Maxillarieae and Cymbidieae, though I have no clear proof of such a relationship.

REFERENCES   Bosser, 1976 (*Hederorkis*); Goss, 1977 (pollination); Kränzlin, 1926 (taxonomy of *Polystachya*, unsatisfactory).

Tribe VANDEAE Lindley

DISCUSSION   The monopodial habit of growth has evolved independently in several different groups, especially in dwarf twig epiphytes, but the Vandeae is the only major group of orchids in which all members are monopodial. In this group the monopodial habit must have been fixed at an early stage of evolution. This habit permits a great flexibility in some respects, but does not permit the development of pseudobulbs. Thus, the leaves of this group may be very fleshy, and the thickest aerial roots of any orchid group are to be found here. Aside from the uniformly monopodial habit, this group is characterized by a distinctly dorsoventral protocorm and relatively uniform chromosome numbers.

RELATIONSHIPS   I have suggested that this tribe may be more closely allied to the Polystachyeae than to other groups. One can easily imagine the evolution of monopodial habit by the production of lateral inflorescences and the retention of apical growth in something similar to *Hederorkis* or *Neobenthamia*, but it is equally possible that the Vandeae do not have any close living relatives.

PHYLETIC TRENDS   In vegetative features, we find condensation of the stem to produce plants similar to *Phalaenopsis* and the evolution of very fleshy leaves, sometimes cylindrical or laterally flattened. The most striking trend is the evolution of leafless plants with photosynthetic roots. We find this trend in all three subtribes, in two or three different groups of Sarcanthinae, in the Angraecinae, and apparently two or more times in the Aerangidinae. Johansson (1977) has suggested that these leafless plants may be partially indirect parasites. In floral features, some trends are clear, but the overall patterns of evolution are not well worked out. I believe that four pollinia is the primitive condition, and that reduction to two pollinia has occurred several times within this tribe. Also, especially in the Angraecinae and Aerangidinae, we find the evolution of two separate stipes and viscidia, each pollinium then having its own stipe and viscidium. In some of the small-flowered groups we find the evolution of four equal spheroid or obovoid pollinia, and there seems to be at least one case of reduction to two pollinia by the loss of one pair, rather than by fusion (Holttum, 1959). One clear trend of floral evolution is the development of a prominent spur. In the Angraecinae and Aerangidinae this is usually associated with the moth-pollination syndrome, and we find parallel development of this syndrome in *Amesiella* and *Neofinetia*, of the Sarcanthinae. There are doubtless other trends in floral evolution, but it is not at all clear what is the primitive flower structure for this tribe, and it is thus difficult to delineate trends.

Holttum has proposed an interesting outline of the evolution of the Vandeae. He considers the possession of two cleft pollinia as primitive, but I agree with Garay (1972b) that reduction from four to two is more likely, and this requires only a slight change in Holttum's scheme (fig. 10.2). Holttum suggests that the large-flowered genera near *Vanda*, *Vandopsis*, and *Arachnis* may be close to the basic pattern for this tribe. These genera are fairly easy to cross, while the small-flowered genera which Holttum considers derivative are, in some cases at least, intersterile. This is a curious pattern, but, of course, we have little data on interfertility in the small-flowered genera. Still, we should also consider the possibility that some small-flowered groups are primitive, and that the large-flowered complex represents only one of several lines of evolution within the group.

REFERENCE   Holttum, 1959 (evolutionary scheme).

Subtribe SARCANTHINAE Bentham

DESCRIPTION   Habit: monopodial, stem short to elongate. Leaves: distichous, rarely secondarily spiral, duplicate, sometimes cylindric, laterally flattened, or lacking. Inflorescence: lateral, simple or branched,

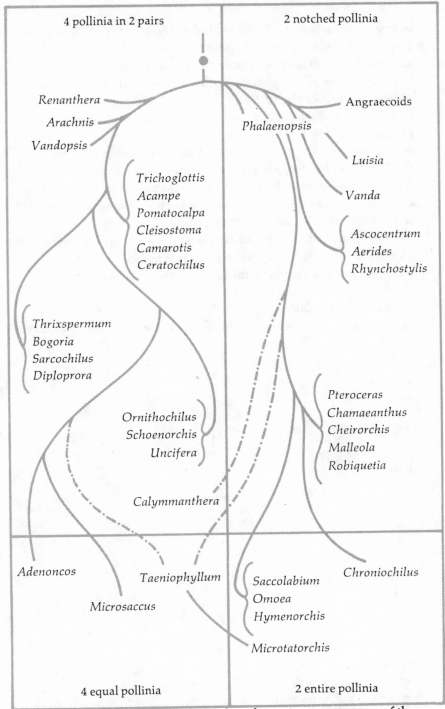

Figure 10.2 A tentative scheme of relationship of some important genera of the Vandeae, arranged according to their pollinaria. (After Holttum, 1959.)

of one to many flowers, flowers spiral or sometimes distichous. Flowers: tiny to very large, lip may be jointed, saccate, or deeply spurred; column may have a prominent foot; anther terminal, operculate, with reduced partitions; four or two pollinia, with a definite stipe and viscidium; stigma entire. (See fig. 10.3.)

DISTRIBUTION   Tropical Asia, with a few species of *Acampe* and *Taeniophyllum* reaching Africa.

POLLINATION   We have one record of carpenter-bee pollination in *Phalaenopsis*, and two for *Vanda*. We may guess that *Amesiella* and *Neofinetia* are moth pollinated, and that *Ascocentrum* and possibly *Renanthera* are bird pollinated, but we have only these three actual records for this huge and complex group.

CHROMOSOME NUMBERS   38, 40. There is one seemingly reliable count of 24 for *Taeniophyllum*, suggesting that this micro-orchid may show reduction in chromosome number, analogous to that reported for *Psygmorchis* (Oncidiinae).

SPECIES   About 1,000.

GENERA   86 in three tentative alliances: (1) *Aerides, Ascochilus, Bier-*

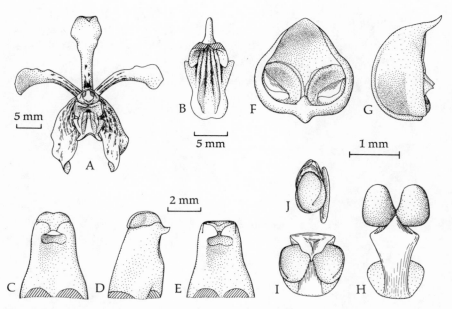

Figure 10.3   *Vanda lamellata* (Vandoideae: Vandeae). *(A)* Flower, front view. *(B)* Lip and spur. *(C)* Column, ventral view. *(D)* Column, side view. *(E)* Column, with anther and pollinarium removed. *(F)* Anther, ventral view. *(G)* Anther, side view. *(H)* Pollinarium, freshly removed. *(I)* Pollinarium, after stipe has curled, top view. *(J)* Pollinarium, after stipe has curled, side view.

mannia, Bogoria, Calymmanthera, Chamaeanthus, Cheirorchis, Chilo-schista, Chroniochilus, Cordiglottis, Cryptopylos, Doritis, Drymoanthus, Grosourdya, Kingidium, Macropodanthus, Ornithochilus, Papillalabium, Peristeranthus, Phalaenopsis, Pteroceras, Rhinerrhiza, Rhynchostylis, Sarcochilus, Sedirea, Seidenfadenia, Thrixsperum. (2) *Adenoncos, Arachnis, Armodorum, Ascocentrum, Ascoglossum, Cleisocentron, Cottonia, Dimorphorchis, Diploprora, Esmeralda, Luisia, Papilionanthe, Paraphalaenopsis, Renanthera, Renantherella, Smitinandia, Vanda, Vandopsis.* (3) *Abdominea, Acampe, Amesiella, Ascochilopsis, Brachypeza, Ceratochilus, Cleisomeria, Cleisostoma, Diplocentrum, Dryadorchis, Eparmatostigma, Gastrochilus, Holcoglossum, Hymenorchis, Loxoma, Malleola, Micropera, Microsaccus, Microtatorchis, Mobilabium, Neofinetia, Omoea, Pelatantheria, Pennilabium, Phragmorchis, Plectorhiza, Pomatocalpa, Porrorhachis, Porphyrodesme, Robiquetia, Saccolabiopsis, Saccolabium, Staurochilus, Stauropsis, Stereochilus, Taeniophyllum, Trachoma, Trichoglottis, Tuberolabium, Uncifera, Ventricularia, Xenikophyton.*

INTERGENERIC HYBRIDS   I know of no natural intergeneric hybrids in this subtribe, but there are many intergeneric hybrids created artificially (see fig. 10.4). This is also the group in which most new intergeneric hybrids are being registered. So far, hybridists have concentrated on the *Vanda* complex, with the *Aerides* complex also receiving a good deal of attention. It is curious that no intergeneric hybrids are registered between genera of the *Acampe* complex, but a number are recorded between the genera of this complex and those of the *Aerides* and *Vanda* alliances. This is probably an artifact reflecting the hybridists' interest in larger flowers.

There are now hybrids registered between *Aerangis* and *Aeranthes*, and between these genera and *Cyrtorchis* and *Angraecum*, respectively. More surprising are crosses between *Eurychone* and *Phalaenopsis* and between *Angraecum* and *Ascocentrum*.

DISCUSSION   This is a large and complex group, and unfortunately no adequate review is available. Workers at Kew have used several subtribal names for this group, and the separation into three alliances used here is based on such a list, but I cannot find that these groups are explained anywhere. One student of orchid taxonomy refers to the Sarcanthinae as "the black pit." One needs to spend years working with this group in order to know it well enough to say anything really meaningful. As an outsider who has not spent the requisite time studying these Asiatic orchids, my first reaction is that the genera may be too finely split. Garay (1972b) justifies the classification by the importance of the floral details

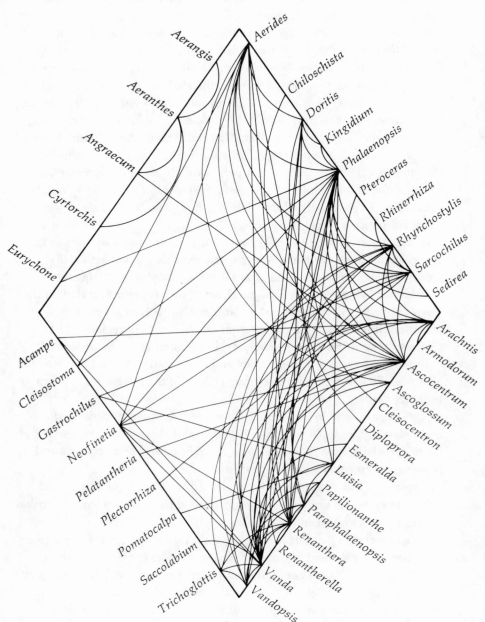

Figure 10.4 Intergeneric hybrids known in the Vandeae. All these are artificial
   hybrids.

in pollination, but we really know next to nothing about pollination, so we can only guess which details are critical. Schlechter (1926) gave great importance to the presence or absence of a column foot but was careless in his observations. More recent authors stress number of pollinia and details of lip structure.

In some ways the Sarcanthinae are reminiscent of the Orchideae, and a thorough-going revision may reduce the number of genera somewhat. In the past, with the emphasis on floral details, species with cylindrical leaves were included in *Aerides, Phalaenopsis,* and *Vanda*. In each case, the cylindrical-leaved species crossed easily with the cylindrical-leaved species of the other genera, but not, or only with difficulty, with their supposed congeners. Now, the cylindrical-leaved *Phalaenopsis* are treated as *Paraphalaenopsis*, and the cylindrical-leaved species of *Vanda* and *Aerides* are united as *Papilionanthe* by Garay (1974b). The impression of excessive splitting is also enhanced by the great number of artificial intergeneric hybrids, but Holttum (1952a) assures us that most of these intergeneric hybrids are sterile; so the classification may not be as bad as it seems at first glance. The systematic arrangement and key promised by Garay (1972b) are eagerly awaited.

REFERENCES Garay, 1972b, 1974b (generic realignments); Holttum, 1952a (hybridization), 1959 (evolutionary trends); Seidenfaden, 1971 (*Luisia*), 1975b (*Cleisostoma* in Thailand); Sweet, 1968-69 (revision of *Phalaenopsis*); Tan, 1975-76 (*Arachnis* and allied genera).

Subtribe ANGRAECINAE Summerhayes

DESCRIPTION Habit: monopodial, stem short to elongate. Leaves: distichous, duplicate, sometimes cylindrical, laterally flattened or lacking. Inflorescence: lateral, simple, of one to many flowers, flowers spiral, secund, or sometimes distichous(?). Flowers: tiny to large, resupinate or not; lip deeply saccate to spurred; column short, anther terminal, operculate, with reduced partitions; two pollinia, with one or two stipes and viscidia; stigma entire, rostellum deeply notched.

DISTRIBUTION Tropical Africa and tropical America.

POLLINATION The moth-pollination syndrome predominates in this group, but we have no observations of pollination. This and the following group could be studied by attracting moths to a "black light" and checking for pollinaria on their tongues.

CHROMOSOME NUMBERS 38, 40, 42.

SPECIES About 400.

GENERA 16 (1) African genera: *Aeranthes, Ambrella, Angraecum, Bonniera, Cryptopus, Jumellea, Lemurella, Neobathiea, Oeonia, Oeoniella,*

Perrierella, Sobennikoffia. (2) American genera: *Campylocentrum, Dendrophylax, Harrisella, Polyradicion.*

DISCUSSION This is primarily a Madagascan group, with some representatives on the mainland of Africa and outliers in the Americas. *Campylocentrum* includes both leafy and leafless species, while the other American genera are all leafless. This pattern suggests that the invader from Africa was a leafy, small-flowered plant, and that the large-flowered species have evolved in the West Indies, where the evolutionary opportunities were evidently different from those on the mainland.

REFERENCES Jones, 1967 (chromosome numbers); Stewart, 1976 (general); Summerhayes, 1966 (delineation of subtribe).

Subtribe AERANGIDINAE Summerhayes

DESCRIPTION Habit: monopodial, stem short to elongate. Leaves: distichous, duplicate, sometimes cylindrical or laterally flattened, or lacking, articulate. Inflorescence: lateral, simple, of one to many flowers, spiral, secund, or distichous(?). Flowers: tiny to large, resupinate; lip deeply saccate or spurred; column short, anther terminal, operculate, two pollinia with one or two stipes and viscidia; stigma entire, rostellum long-beaked. (See fig. 10.5.)

DISTRIBUTION Tropical Africa.

POLLINATION Johansson (1974) indicates a moth, *Euchromia*, as a frequent visitor of *Diaphananthe*, and moth pollination is to be expected in many angraecoids. Johansson also lists chrysomelid beetles as frequent visitors of *Cyrtorchis* and *Tridactyle*, but these may not be pollinators.

CHROMOSOME NUMBERS 46, 50. Fifty is the predominant number.

SPECIES About 300.

GENERA 34: *Aerangis, Ancistrorhynchus, Angraecopsis, Barombia, Beclardia, Bolusiella, Calyptrochilum, Cardiochilus, Chamaeangis, Chauliodon, Cyrtorchis, Diaphananthe, Dinklageella, Distylodon, Eggelingia, Encheiridion, Eurychone, Lemurorchis, Listrostachys, Microcoelia, Mystacidium, Nephrangis, Plectrelminthus, Podangis, Rangaeris, Rhaesteria, Rhipidoglossum, Solenangis, Sphyrarhynchus, Summerhayesia, Taeniorrhiza, Triceratorhynchus, Tridactyle, Ypsilopus.*

DISCUSSION Though this group is very close to the Angraecinae, it seems to be well characterized by both floral features and chromosome number. While the Angraecinae are better developed on Madagascar, the Aerangidinae are much better developed on mainland Africa.

REFERENCES Stewart, 1976 (general); Summerhayes, 1966 (delineation of subtribe).

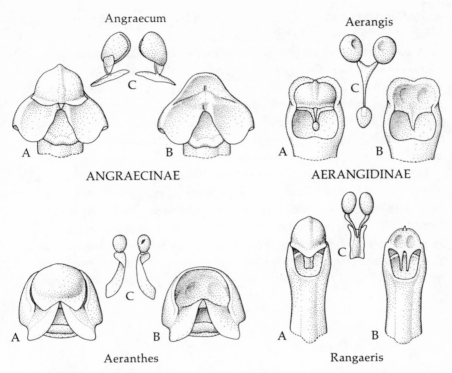

Angraecum

Aerangis

ANGRAECINAE

AERANGIDINAE

Aeranthes

Rangaeris

Figure 10.5 A comparison of Angraecinae and Aerangidinae. *(A)* Column, with anther in place. *(B)* Column, with anther and pollinarium removed. *(C)* Pollinarium (or hemipollinaria). (After Stewart, 1974.)

Tribe MAXILLARIEAE Pfitzer

DISCUSSION   This group includes all the vandoid orchids with four pollinia except the Vandeae and the Polystachyeae, and seems to be a rather natural group as here constituted. It is primarily an American group, but the most primitive subtribe, the Corallorhizinae, is represented in Eurasia as well as in the Americas. The primitive groups are basically cormous with plicate leaves, and this contrasts with the Polystachyeae, which have rather primitive features associated with reedlike stems and conduplicate leaves. In the more advanced groups we find a good deal of diversity in all features. The pattern of variation within this tribe is rather puzzling. We find only one-flowered inflorescences in the Maxillariinae and the Dichaeinae, and some genera with one-flowered inflorescences are found in the Zygopetalinae, Bifrenariinae, and Lycastinae. Other features suggest, however, that the genera with one-flowered inflorescences are not very closely related between the subtribes. There is no doubt that all these subtribes are closely interrelated, though, and we may eventually find that the whole system needs to be changed. (See fig. 10.6.)

RELATIONSHIPS   This tribe is probably more closely allied to the Cymbidieae than to the other vandoid tribes.

PHYLETIC TRENDS   Vegetatively we find the trends characteristic of the family, from corms of several internodes and plicate leaves to pseudobulbs of a single internode and conduplicate leaves. In some groups we find monopodial growth, and cylindrical or laterally flattened leaves also occur.

Florally, the group shows less extreme diversity. Spurs are found in *Tipularia* and in a few Maxillariinae, but are infrequent. Most of the *Zygopetalum* complex seems to have adapted to pollination by euglossine bees, but have not evolved such bizarre mechanisms as in the Stanhopeinae. Some Telipogoninae (perhaps all) have evolved pseudocopulation, while the Ornithocephalinae, or at least some of them, produce oil and are pollinated by *Paratetrapedia* and possibly by other oil-gathering

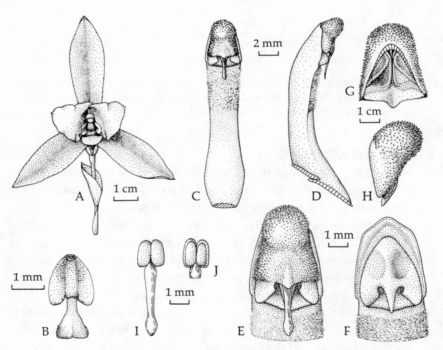

Figure 10.6   *Lycaste tricolor* (Vandoideae: Maxillarieae). *(A)* Flower, front view. *(B)* Lip, flattened. *(C)* Column, ventral view. *(D)* Column, side view. *(E)* Apex of column, with anther in place. *(F)* Apex of column, with anther and pollinarium removed. *(G)* Anther, ventral view. *(H)* Anther, side view. *(I)* Pollinarium. *(J)* Pollinia, ventral view.

bees as well. In the Ornithocephalinae we find a greater degree of floral diversity, and we find the evolution of four equal and spheroid pollinia.

Subtribe CORALLORHIZINAE Camus, Bergan and Camus

DESCRIPTION   Habit: terrestrial, or saprophytic, with corms of several internodes. Leaves: convolute, plicate, articulate, sometimes petiolate. Inflorescence: lateral, simple, of few to many spiral flowers. Flowers: small to medium, lip with or without a spur; column may have a prominent foot; anther terminal, operculate, with poorly developed partitions; four pollinia, superposed, with a distinct viscidium and sometimes a small stipe; stigma entire.

DISTRIBUTION   North temperate and tropical America.

POLLINATION   The genus *Tipularia* is pollinated by noctuid moths. The asymmetry of the flowers permits the deposition of the pollinaria on the compound eye, which is probably the only place on the whole scaly insect where the viscidia could stick well. Syrphid fly pollination has been reported for *Corallorhiza*.

CHROMOSOME NUMBERS   36, 40, 42, 46, 48, 50.

SPECIES   About 60.

GENERA   9: *Aplectrum, Corallorhiza, Cremastra, Dactylostalix*(?), *Didiciea, Ephippianthus, Govenia, Oreorchis, Tipularia*.

DISCUSSION   These genera have been placed in quite diverse groups in previous classifications. *Corallorhiza* has been associated with the very different *Hexalectris,* though its flower structure is virtually identical to that of *Oreorchis*. Some genera have been associated with the Cyrtopodiinae, though these genera have four rather than two pollinia, and they are distinct in aspect as well. Several genera have been associated with the Malaxideae, but these have superposed pollinia, and I believe that adequate material will show that all have a viscidium, and possibly a stipe. *Govenia*, the only tropical American genus assigned here, is distinctive in habit and flower structure, and may be only distantly related to the other genera. It is possible that some of the Asiatic genera, such as *Dactylostalix*, will prove to belong with the Calypsoeae when details of column structure and development are known.

REFERENCE   Stoutamire, 1978 (pollination of *Tipularia*).

Subtribe ZYGOPETALINAE Schlechter

DESCRIPTION   Habit: terrestrial or epiphytic, with pseudobulbs of several internodes or of one internode, or stems slender, short or rarely elongate. Leaves: convolute or duplicate, plicate or conduplicate, usually distichous, articulate. Inflorescence: lateral, of one or several spiral flowers, often from a young growth. Flowers: small to large, resupinate or not; column short or elongate, often winged or flattened, usually with a distinct foot; anther terminal or rather ventral, operculate, with reduced partitions; four pollinia, superposed, with a prominent viscidium and usually with a well-developed stipe; stigma entire.

DISTRIBUTION   Tropical America.

POLLINATION   All of our records indicate that these orchids are pollinated by euglossine bees. In the simplest cases, they function as gullet flowers and place the pollinaria on the back of the bee, or especially behind the head. In the *Chondrorhyncha* complex, however, we find a few complexities, one being that we have collected female eulaemas with the pollinaria of *Cochleanthes*. We do not know what attracts the females to the flowers. *Kefersteinia* places the pollinaria on the basal segment (scape) of the bee's antenna, while *Chaubardiella* places the pollinaria on the trochanter of the bee's leg.

CHROMOSOME NUMBERS   46, 48 (very few counts are known).

SPECIES   About 150.

GENERA   26 in four closely related alliances: (1) Corms or pseudobulbs of several internodes: *Otostylis, Warrea, Warreella, Warreopsis*. (2) Pseudobulbs of a single internode or stems elongate, inflorescence usually of several flowers: *Aganisia, Batemannia, Cheiradenia, Koellensteinia, Mendoncella, Neogardneria, Pabstia, Paradisianthus, Promenaea, Zygopetalum, Zygosepalum*. (3) Pseudobulbs reduced or absent, inflorescence one-flowered: *Bollea, Chaubardia, Chaubardiella, Chondrorhyncha, Cochleanthes, Hoehneella, Huntleya, Kefersteinia, Pescatorea, Stenia*. (4) Slender, monopodial plant, relationships uncertain: *Vargasiella*.

INTERGENERIC HYBRIDS   One natural intergeneric hybrid is recorded between *Pabstia* and *Zygopetalum*, and there are several natural hybrids between *Bollea* and *Pescatorea*. Though this subtribe has received relatively little attention from hybridizers, most genera appear to be crossable (see fig. 10.7). At one time, many intergeneric hybrids of *Zygopetalum* were mentioned in the literature, but most of these were the offspring of an apomictic strain of *Zygopetalum*. The plants would produce seed when the stigma was stimulated by foreign pollinia, but the seedlings were all *Zygopetalum*. However, hybrids are recorded between *Zygopetalum* and *Lycaste* and, as far as I know, these hybrids were authentic, but few crosses have been registered of this combination.

DISCUSSION   This group is somewhat diverse in terms of key features, but it is held together by both interfertility and close resemblances. The first series, with plicate leaves, corms or pseudobulbs of several internodes, and terminal anthers, seems very like the Corallorhizinae and

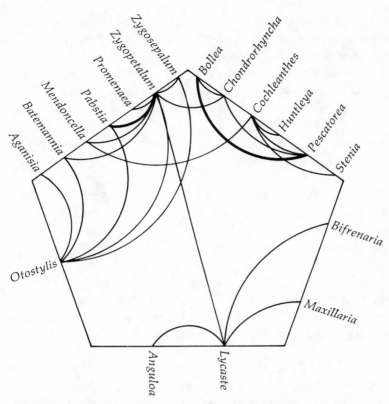

Figure 10.7 Intergeneric hybrids known in the Maxillarieae. The heavy lines indicate naturally occurring hybrids.

surely represents the primitive element of this subtribe. There is no sharp line between this and the second series, in which *Mendoncella*, *Zygopetalum*, and *Zygosepalum* have the anther ventral on the column. The third series, with very reduced pseudobulbs (or none), more or less conduplicate leaves, and one-flowered inflorescences, is a distinctive and easily recognized group, but closely allied to the preceding. The Andean *Vargasiella* has been described as having naked pollinia, but the pollinia appear to be superposed, and I suspect that it is closer to this group than to any other. Better material of this puzzling genus is needed.

REFERENCES   Garay, 1969 (*Chondrorhyncha* complex), 1973 (*Zygopetalum* complex).

Subtribe BIFRENARIINAE Dressler, 1979b

DESCRIPTION   Habit: epiphytic or lithophytic, with pseudobulbs of a single internode, these sometimes covered with hard sheaths. Leaves:

convolute, plicate or subconduplicate, terminal or distichous, articulate. Inflorescence: lateral, of one to several spiral flowers. Flowers: small to large, usually resupinate, column short or elongate, usually with a prominent foot; anther terminal, operculate, with reduced partitions; four pollinia, superposed, with a prominent viscidium, the pollinia sessile on the viscidium or with one or usually two stipes; stigma entire.

DISTRIBUTION   Tropical America.

POLLINATION   We have found pollinaria of *Bifrenaria* on males of *Eufriesea violacea* (Euglossini), but this is our only information on this genus. There are two observations of *Xylobium* being pollinated by stingless bees *(Trigona)*. The floral resemblances between *Xylobium* and *Maxillaria* may be a convergence caused by the pollination system.

CHROMOSOME NUMBERS   38, 40.

SPECIES   About 50.

GENERA   5: *Bifrenaria, Horvatia, Rudolfiella, Teuscheria, Xylobium.*

INTERGENERIC HYBRIDS   Crosses between *Bifrenaria* and *Lycaste* are registered, but very few are known. I have no information on their fertility.

DISCUSSION   This small complex shows definite relationships to the Lycastinae and the Zygopetalinae but does not fit well in either one.

Subtribe LYCASTINAE Schlechter

DESCRIPTION   Habit: epiphytic or terrestrial, with pseudobulbs of a single internode. Leaves: convolute, plicate, distichous or terminal, articulate. Inflorescence: lateral, of one or many spiral flowers. Flowers: medium to large, resupinate or erect; column elongate, with a prominent foot; anther terminal, operculate, with reduced partitions; four pollinia, superposed, with a prominent viscidium and a long, narrow stipe; stigma entire.

DISTRIBUTION   Tropical America.

POLLINATION   Both *Anguloa* and *Lycaste* are pollinated by euglossine bees, but the pollinaria of *Lycaste* are sometimes found on female bees, suggesting that the flowers offer some reward other than perfume. *Anguloa*, however, is typical of the euglossine-pollination syndrome and attracts only male bees, which are flipped against the column by the hinged lip.

CHROMOSOME NUMBER   40.

SPECIES   About 40.

GENERA   3: *Anguloa, Lycaste, Neomoorea.*

INTERGENERIC HYBRIDS   There are a number of artificial hybrids between *Auguloa* and *Lycaste,* and a few crosses are registered between *Lycaste* and *Bifrenaria, Zygopetalum* and *Maxillaria.*

DISCUSSION The genus *Neomoorea* has been classified with the Stanhopeinae, but it is clearly out of place there. Its pollinarium is very like
that of *Lycaste*, but the inflorescence has many flowers.
REFERENCE Fowlie, 1970 (revision of *Lycaste*).

Subtribe MAXILLARIINAE Bentham

DESCRIPTION Habit: epiphytic or terrestrial, with pseudobulbs of one
internode, or stem slender, short or elongate. Leaves: distichous or secondarily spiral, duplicate, articulate. Inflorescence: lateral, of a single
flower. Flowers: small to large, lip usually hinged, sometimes forming
a saccate nectary with the column foot, may occasionally have a distinct
spur; column slender or short, anther terminal, operculate, with reduced
partitions, four pollinia, superposed, with viscidium and more or less
well-developed stipe; stigma entire.
DISTRIBUTION Tropical America.
POLLINATION The available records indicate pollination by various food-
seeking bees for *Maxillaria*. Some Brazilian species are evidently pollinated by bees that gather wax from the lip. Braga (1978) reports that
Maxillaria pendens is pollinated by *Stelopolybia*, a vespid wasp, which is
apparently gathering wax from the lip. *Cryptocentrum* strongly suggests
moth pollination, with its dull, deeply spurred flowers.
CHROMOSOME NUMBERS 40, 48.
SPECIES About 485.
GENERA 9: *Anthosiphon, Chrysocycnis, Cryptocentrum, Cyrtidium,
Maxillaria, Mormolyca, Pityphyllum, Scuticaria, Trigonidium.*
INTERGENERIC HYBRIDS A number of crosses between *Maxillaria* and
Lycaste have been attempted, and seedlings have been obtained in
several cases. Two such crosses have been registered.
DISCUSSION While related to the Zygopetalinae, this group seems distinctive in its clearly conduplicate leaves and one-flowered inflorescence.
Cryptocentrum is distinctive in its long-spurred flowers but seems to
belong here. Brieger (1977) discusses *Cryptocentrum* and other genera
with prominent nectaries and treats both *Sepalosaccus* and *Pseudomaxillaria* as valid genera. The whole *"Ornithidium"* complex needs
careful study. Many species show the fixed lip of *Pseudomaxillaria* in
greater or lesser degree, and *Sepalosaccus* seems but one of the complex.
REFERENCE Brieger, 1977 (genera with prominent nectaries).

Subtribe DICHAEINAE Schlechter

DESCRIPTION Habit: monopodial. Leaves: distichous, duplicate, articulate
or not. Inflorescence: lateral, arising opposite from the leaf axil, one-

flowered. Flowers: small to medium-small, resupinate; column short, may have a slight foot; anther terminal, operculate, with reduced partitions; four pollinia, superposed, with a distinct viscidium and stipe; stigma entire.

DISTRIBUTION   Tropical America.

POLLINATION   We have a few observations of pollination by euglossine bees, and we frequently see *Dichaea* pollinaria on the clypeus (face) of bees, but these pollinaria are very delicate and are usually lost in the net or the killing jar.

CHROMOSOME NUMBER   52.

SPECIES   About 45.

GENUS   *Dichaea.*

DISCUSSION   This distinctive genus reminds one of the *Chondrorhyncha* complex and may be derived from Zygopetaline-like ancestors. The leaf-opposed inflorescence is puzzling, and the habit may really be extreme sympodial branching, with each inflorescence terminal on a stem of one internode. The group is poorly known, and the number of species could easily be twice that given.

REFERENCE   Kränzlin, 1923 (revision, inadequate).

Subtribe TELIPOGONINAE Schlechter

DESCRIPTION   Habit: epiphytic, with pseudobulbs of one internode, or slender stems. Leaves: distichous, duplicate, articulate, occasionally leafless. Inflorescence: terminal or lateral, of few to many flowers, spiral or distichous. Flowers: tiny to medium-large, resupinate or erect; column short, usually bristly; anther dorsal, erect, with reduced partitions; four pollinia, superposed, with a long stipe and a hooked viscidium; stigma entire.

DISTRIBUTION   Tropical America.

POLLINATION   *Trichoceros* is pollinated through pseudocopulation by bristly tachinid flies, and similar flies have been observed pollinating *Telipogon* (Andrés Maduro, pers. comm.).

CHROMOSOME NUMBER   Not known.

SPECIES   About 60.

GENERA   4: *Dipterostele, Stellilabium Telipogon, Trichoceros.*

DISCUSSION   *Telipogon* is unusual in the possession of a terminal inflorescence, an unexpected feature in an advanced member of this tribe. *Stellilabium* approaches the leafless condition of some Vandeae, often being leafless at flowering, but with a flattened green inflorescence in addition to the green roots.

Subtribe ORNITHOCEPHALINAE Schlechter

DESCRIPTION Habit: dwarf epiphytes, with or without pseudobulbs, stem short, sympodial or monopodial. Leaves: distichous or secondarily spiral, duplicate, may be laterally flattened, articulate or not. Inflorescence lateral, simple, of few to many flowers, flowers spiral, secund, or(?) distichous. Flowers: small, lip with a prominent oil gland, or deeply saccate to somewhat spurred; column slender; anther terminal, operculate, beaked, with reduced partitions; four pollinia, superposed or obovoid, with a viscidium and a long stipe; stigma tends to be basal on the column, entire; rostellum beaked, short to very long.

DISTRIBUTION Tropical America.

POLLINATION We have several observations of *Ornithocephalus* being pollinated by bees of the genus *Paratetrapedia*. Vogel (1974) has indicated that the gland on the lip is actually an oil gland.

CHROMOSOME NUMBER Not known.

SPECIES About 70.

GENERA 14: *Centroglossa, Chytroglossa, Dipteranthus, Dunstervillea, Eloyella, Hintonella, Hofmeisterella, Ornithocephalus, Phymatidium, Platyrhiza, Rauhiella, Sphyrastylis, Thysanoglossa, Zygostates.*

DISCUSSION These little micro-orchids may be related to the Telipogoninae, but such a relationship is tentative. I would guess that *Hintonella* and *Dunstervillea* are among the more primitive members of the group, while *Ornithocephalus* and *Sphyrastylis* are quite specialized.

Tribe CYMBIDIEAE Pfitzer

DISCUSSION In this group I place all vandoid orchids with two pollinia except the Vandeae. The potential weakness of this system is obvious. We know that the reduction from four to two pollinia has occurred at least twice within the Vandeae. How can we be sure that it has not occurred two or three times to give rise to the orchids that I group together here? At present we cannot be sure, and I have considered giving tribal status to each of the last three subtribes. Still, I am not at all sure that the group is polyphyletic. The most dubious groups, surely, are the Stanhopeinae and the Pachyphyllinae, neither of which shows clear ties to any other group. The Catasetinae clearly belong with the Cyrtopodiinae, and I believe that the same is true of the Oncidiinae, even though they are one of the most distinctive of all orchid subtribes.

RELATIONSHIPS I believe that the Cymbidieae share a common ancestry with the Maxillarieae, and more specifically with the Corallorhizinae, which must be very like the common ancestor of both tribes.

PHYLETIC TRENDS The vegetative trends shown here are much the same as for the Maxillarieae. Florally, however, the pattern is different. In the Cyrtopodiinae some genera show a fairly simple gullet flower, while others have well-developed spurs. In the Catasetinae we find the development of an elastic stipe and violent discharge of the viscidium, and with this we find the evolution of unisexual flowers, which are found in no other orchid group. In the Stanhopeinae we also have simple gullet flowers, but the evolution of euglossine pollination has permitted an impressive array of unlikely mechanisms which I will treat at greater length under that group. The Pachyphyllinae is a minor group, but bird pollination is clearly the factor behind the evolution of *Fernandezia*. In the Oncidiinae, I believe that the presence of a spur is the primitive condition and that the spur has been lost in most groups of Oncidiinae as they have evolved euglossine pollination, pollination by anthophorid bees, or the deceit of other large bees.

Subtribe CYRTOPODIINAE Bentham

DESCRIPTION   Habit: terrestrial, epiphytic, or occasionally saprophytic, with corm or pseudobulb of several internodes, rarely with slender reed-like stems or pseudobulbs of one internode. Leaves: spiral or distichous, convolute or duplicate, plicate or conduplicate, articulate, often scattered along pseudobulb. Inflorescence: lateral or occasionally terminal, simple or branched, of few to many spiral flowers. Flowers: small to large, resupinate, lip may be saccate or deeply spurred, or hinged; column usually with a prominent foot, may have wings; anther terminal, operculate, with reduced partitions; two pollinia, notched or cleft, with viscidium and usually with stipe; stigma entire. (See fig. 10.8.)

DISTRIBUTION   Pantropical.

POLLINATION   Pollination by carpenter bees has been observed in *Eulophia* and *Cymbidium*, while pollination by *Apis* and a wasp are also reported for *Cymbidium*. Pollination by *Euglossa* is reported for *Cyrtopodium*, and the pollinaria of *Galeandra* have been found on *Euglossa* and large anthophorid bees. Probably most members of this subtribe are pollinated by bees of appropriate sizes.

CHROMOSOME NUMBERS   32, 36, 38, 40, 42, 54, 56.

SPECIES   About 425.

GENERA   24 in five tentative alliances: (1) *Bromheadia, Claderia*. (2) *Chrysoglossum, Cyanaeorchis, Diglyphosa, Eulophia, Eulophiella, Geodorum, Oecoclades, Pteroglossaspis*. (3) *Acrolophia, Cymbidiella, Cyrtopodium, Eriopsis, Galeandra, Grammangis, Graphorkis, Grobya*. (4) *Ansellia, Cymbidium, Grammatophyllum, Poicilanthe(?), Porphyroglottis(?)*. (5) *Dipodium*.

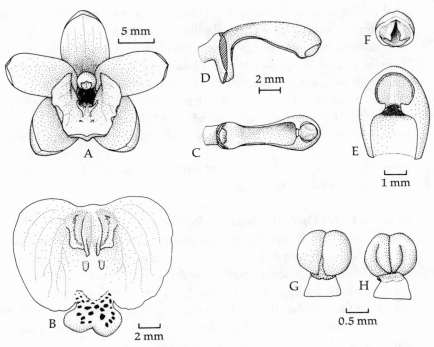

Figure 10.8 *Eriopsis biloba* (Vandoideae: Cymbidieae). *(A)* Flower, front view. *(B)* Lip, flattened. *(C)* Column, ventral view. *(D)* Column, side view. *(E)* Apex of column, with anther and pollinarium removed. *(F)* Anther. *(G)* Pollinarium, dorsal view. *(H)* Pollinarium, ventral view.

INTERGENERIC HYBRIDS   The genus *Cymbidium* has been crossed success-fully with *Ansellia* and *Grammatophyllum*, though the hybrids are re-putedly slow to grow and flower. Attempts to cross *Cymbidium* with *Grammangis* and other genera have not been successful.

DISCUSSION   This is a very diverse group, but none of the systems of division has been entirely satisfactory. The tentative division into alliances which I give here may be no more satisfactory but will serve to emphasize the diversity (without adding any new names). *Brom-headia* and *Claderia* have slender, reedlike stems and thus seem out of place in this primarily cormous group, and each is different from the other. *Ansellia* and *Grammatophyllum* both have the same or nearly the same chromosome number as *Cymbidium*, and both genera have been crossed with *Cymbidium*. Other genera which are similar in appearance have quite different chromosome numbers and apparently cannot be crossed with *Cymbidium*. Thus, I have tentatively grouped *Grammangis* and *Cymbidiella* with the *Cyrtopodium* complex, even though they resemble *Cymbidium* in general structure. The members of the *Cyrtopodium* complex usually have pseudobulbs and, as far as known, higher chromosome numbers, but the distinction between this and the primarily cormous *Eulophia* complex may not hold up. *Dipodium* is customarily placed with *Cymbidium*, but its flower structure is very different, and Clifford and Smith (1969) suggest that its seed features are more like those of the *Cyrtopodium* and *Eulophia* complexes; *Dipodium* has prominent, hyaline caudicles that are quite unlike those of any of the allied genera.

REFERENCES   Garay and Taylor, 1976 *(Oecoclades)*; Lock and Profita, 1975 (pollination of *Eulophia*).

Subtribe GENYORCHIDINAE Schlechter

DESCRIPTION   Habit: epiphytic, with pseudobulbs of one internode scattered on a long rhizome. Leaves: terminal on pseudobulb, duplicate, articulate. Inflorescence: lateral, of several spiral flowers. Flowers: small, nonresupinate, lip hinged, three-lobed; column wingless, short, with a very long foot; anther terminal, operculate, with reduced partitions; two pollinia, subglobose, without grooves, with stipe and viscidium; stigma entire.

DISTRIBUTION   Tropical Africa.

POLLINATION   Not known.

CHROMOSOME NUMBER   Not known.

SPECIES   About 6.

GENUS   *Genyorchis.*

DISCUSSION   This small genus has been associated with the *Bulbophyllum*-like genera that have, or are supposed to have, stipes, but *Genyorchis* is not at all closely allied to those Asiatic genera. The habit is reminiscent of *Bulbophyllum*, and the flowers are superficially very like those of *Polystachya*. The pollinia are only two, however, without groove or notch, and the genus appears to be closest to the *Cyrtopodiinae*. The stipe is easily split in dissection, and this may have lead Schlechter (1926) to compare *Genyorchis* with *Sunipia*.

Subtribe THECOSTELINAE Schlechter

DESCRIPTION   Habit: epiphytic, with pseudobulbs of one internode. Leaves: terminal on pseudobulb, duplicate, articulate. Inflorescense: lateral, of several to many spiral flowers. Flowers: medium-small, nonresupinate, lip deeply saccate, lip and column united; column more or less sigmoid, winged; anther terminal, operculate, with reduced partitions; two pollinia, grooved, with viscidium and a broad stipe; stigma entire.

DISTRIBUTION   Tropical Asia.
POLLINATION   Not known.
CHROMOSOME NUMBER   Not known.
SPECIES   About 5.
GENUS   *Thecostele*.
DISCUSSION   The flower structure of this genus is so bizzare that its subtribal status seems reasonable, even though it shows some resemblance to *Porphyroglottis* in the form of the column.

Subtribe ACRIOPSIDINAE; Dressler, 1979b

DESCRIPTION   Habit: epiphytic, with pseudobulbs of several internodes. Leaves: terminal, duplicate, articulate (above the base). Inflorescence: lateral, simple, of several to many spiral flowers. Flowers: small, resupinate, lip and column highly united, column with armlike wings and hooded clinandrium; anther suberect, with reduced partitions; two pollinia, laterally flattened, bladelike, with a slender stipe and viscidium; stigma entire, narrowly elliptical.
DISTRIBUTION   Tropical Asia.
POLLINATION   Not known.
CHROMOSOME NUMBER   Not known.
SPECIES   About 12.
GENUS   *Acriopsis*.
DISCUSSION   This genus has been included in the Thecostelinae, but there is little relationship between *Acriopsis* and *Thecostele*; and, indeed, *Acriopsis* is the more isolated of the two. It is doubtless related to the Cyrtopodiinae, but I cannot indicate a relationship to any particular genus.

Subtribe CATASETINAE Schlechter

DESCRIPTION   Habit: epiphytic, with pseudobulbs of several internodes. Leaves: distichous, scattered on pseudobulbs, convolute, plicate, articulate. Inflorescence: lateral, simple, of few to many spiral flowers. Flowers: medium-small to rather large, resupinate or not, often unisexual; column usually with an elastic viscidium-throwing device; anther rather ventral, with reduced partitions; two pollinia, with a viscidium and an elastic stipe; stigma entire.
DISTRIBUTION   Tropical America.
POLLINATION   These orchids are always pollinated by male euglossine bees. *Clowesia* is a simple gullet flower, and the trigger mechanism merely causes the viscidium to drop down a bit, where it will catch on the bee's scutellum (in most species). In the other genera, the pollination mechanisms are diverse and more complicated.

CHROMOSOME NUMBERS   54, 64, 68.

SPECIES   About 145.

GENERA   5: *Catasetum, Clowesia, Cycnoches, Dressleria, Mormodes.*

INTERGENERIC HYBRIDS   Artificial hybrids are registered between *Catasetum* and *Clowesia, Cycnoches,* and *Mormodes,* as well as between *Cycnoches* and *Mormodes.* Probably all combinations in this subtribe will prove to be possible.

DISCUSSION   This is a very distinctive group, but it is obviously related to the *Cyrtopodium–Galeandra* alliance of the Cyrtopodiinae.

REFERENCES   Allen, 1952 (revision of *Cycnoches*); Dodson, 1975 (generic classification); Jones and Daker, 1968 (chromosome numbers); Mansfeld, 1932 (taxonomy of *Catasetum*); Pabst, 1978 (key to *Mormodes*).

Subtribe STANHOPEINAE Bentham

DESCRIPTION   Habit: epiphytic, with pseudobulbs usually of one internode. Leaves: terminal, convolute, plicate, articulate, usually petiolate. Inflorescence: lateral, of one to several spiral flowers. Flowers: small to very large, resupinate or not, sometimes pendant; column winged or not; anther terminal or ventral, operculate, with reduced partitions; two pollinia, cleft, with viscidium and usually a prominent stipe; stigma entire.

DISTRIBUTION   Tropical America.

POLLINATION   These orchids are always pollinated by male euglossine bees. In the primitive genera, such as *Acineta* or *Lycomormium,* they are simple gullet flowers, but the other genera have a bewildering array of pollination systems (see fig. 4.10). *Sievekingia* is essentially a gullet flower, but with the bee entering upside down with respect to the column. *Coeliopsis* is a keyhole flower, and *Peristeria* is a flip-trap flower. There are also hinge flowers, in which the weight of the bee causes the column and the lip to come together, as in *Kegeliella, Paphinia,* and *Polycycnis,* and a variety of fall-through flowers, such as *Gongora, Lacaena,* and *Stanhopea.* The fall-in flower of *Coryanthes* is clearly in a class by itself. Most of the genera place pollinaria beneath the bee's scutellum or on the thorax, but *Kegeliella* and *Lacaena* place the pollinia behind the head, and *Coeliopsis* places its pollinaria on the face. *Cirrhaea, Sievekingia,* and probably *Houlletia* section *Neohoulletia* place their pollinia on the bee's legs. *Schlimmia* is a gullet flower, but it is surely secondarily so, having evolved from something like *Trevoria,* perhaps in conjunction with a smaller pollinator (see fig. 10.9.)

CHROMOSOME NUMBERS   38, 40, 42.

SPECIES   About 190.

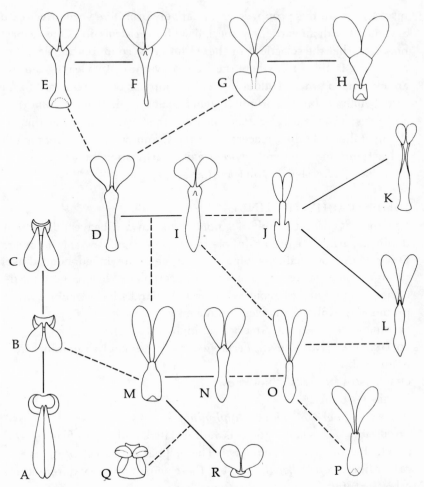

Figure 10.9 Pollinaria of the subtribe Stanhopeinae, arranged according to a tentative scheme of evolution. *(A) Lycomormium. (B) Peristeria. (C) Coeliopsis. (D) Houlletia* section *Neohoulletia. (E) Trevoria. (F) Schlimmia. (G) Stanhopea. (H) Sievekingia. (I) Houlletia* section *Houlletia. (J) Kegeliella. (K) Paphinia. (L) Polycycnis. (M) Acineta. (N) Lacaena. (O) Gongora. (P) Cirrhaea. (Q) Coryanthes. (R) Lueddemannia.* (After Dressler, 1976b.)

GENERA 17: *Acineta, Cirrhaea, Coeliopsis, Coryanthes, Gongora, Houlletia, Kegeliella, Lacaena, Lueddemannia, Lycomormium, Paphinia, Peristeria, Polycycnis, Schlimmia, Sievekingia, Stanhopea, Trevoria.*

INTERGENERIC HYBRIDS Artificial hybrids have been made between *Stanhopea* and both *Acineta* and *Polycycnis*. These genera are rather diverse, and one would guess that many other combinations will prove to be viable.

DISCUSSION   In this group we find relatively primitive vegetative features associated with an extremely specialized flower structure. One suspects, however, that the selection for the evolution of conduplicate leaves is not strong in this case, where there are only one or two leaves and they are terminal on the pseudobulbs. This group has been compared to the Zygopetalinae, but the primitive genera, such as *Acineta, Lycomormium,* and *Peristeria,* seem at least as much like the Cyrtopodiinae as the Zygopetalinae, so their placement in the Cymbidieae seems reasonable, at least until more concrete evidence of their relationships is known.
REFERENCE   Dressler, 1976b (pollinaria).

Subtribe PACHYPHYLLINAE Pfitzer

DESCRIPTION   Habit: epiphytic, monopodial, dwarf. Leaves: distichous, duplicate, articulate or not. Inflorescence: lateral, of few spiral flowers. Flowers: tiny or small, resupinate or not; column winged, hooded over the anther; anther terminal or ventral, operculate, with reduced partitions; two pollinia, obovoid, with viscidium and stipe, sometimes with prominent, hyaline caudicles; stigma entire.
DISTRIBUTION   Tropical America (at high altitudes).
POLLINATION   Not known; *Fernandezia* is presumably hummingbird pollinated.
CHROMOSOME NUMBER   Not known.
SPECIES   About 25.
GENERA   2: *Fernandezia, Pachyphyllum.*
DISCUSSION   This small group is very isolated, and it is difficult to say to which group it may be related. The pollinaria of *Fernandezia* are rather like those of *Lockhartia,* but these genera are very different in other features.
REFERENCES   Kränzlin, 1923 (taxonomy, unsatisfactory).

Subtribe ONCIDIINAE Bentham

DESCRIPTION   Habit: epiphytic or terrestrial, usually with pseudobulbs of a single internode, sometimes with short or elongate, slender stems, sometimes monopodial. Leaves: distichous, duplicate, articulate or not, may be cylindrical or laterally flattened. Inflorescence: lateral, simple or branched, of one to many flowers, flowers spiral or distichous. Flowers: tiny to very large, resupinate or not(?), lip may be spurred, or may have basal nectariferous appendages extending into a sepaline spur, often with large calli; column winged or not, anther terminal or erect and dorsal, operculate, with reduced partitions; two pollinia, with viscidium and stipe, the stipe and viscidium usually appearing different in color or texture; stigma entire or two-lobed.

DISTRIBUTION   Tropical America.

POLLINATION   This group shows a great diversity of pollination systems. We find wasp pollination in *Ada, Brassia,* and *Leochilus.* Both humming-bird pollination and butterfly pollination are reported for *Rodriguezia lanceolata* (van der Pijl and Dodson, 1966; Braga, 1978). Hummingbird pollination is also reported for *Comparettia* and is to be expected in several other genera, including one species of *Ada. Gomesa* and *Miltoniopsis* are bee pollinated, and both *Oncidium* and *Odontoglossum* seem to be bee pollinated. In the case of *Oncidium,* pollination by *Centris* seems to predominate, sometimes motivated by food, sometimes by deceit of one sort or another, but pollination by bumblebees and carpenter bees is reported for two aberrant groups. *Sigmatostalix* has oil glands on the lip and, like some species now included in *Oncidium,* attracts oil-gathering bees, but in this case only very small bees. We know of euglossine pollination in *Aspasia, Lockhartia, Notylia, Rodriguezia, Trichocentrum,* and *Trichopilia.* This does not mean that a large sector of this subtribe is pollinated by euglossine bees, but rather that our sampling of euglossine-pollinated orchids is much better than for other classes. (See fig. 10.10.)

CHROMOSOME NUMBERS   10, 14, 24, 26, 28, 30, 36, 38, 40, 42, 44, 48, 50, 56, 60. Both five and seven (haploid) have been suggested as base numbers for this subtribe, but I am a bit skeptical. This is a very low number for the orchids, and hardly to be expected as a primitive feature in such highly evolved orchids as *Psygmorchis.* It seems likely that the number of *Psygmorchis* is reduced from a higher number, as is known to occur in many short-lived plants.

SPECIES   About 950.

GENERA   57: five alliances (fig. 10.11), of which the first is very diverse: (1) *Ada, Amparoa, Aspasia, Brachtia, Brassia, Capanemia, Caucaea, Cochlioda, Erycina, Gomesa, Hybochilus, Leochilus, Mesospinidium, Mexicoa, Miltonia, Miltoniopsis, Neodryas, Odontoglossum, Oncidium* (most sections), *Ornithophora, Palumbina, Papperitzia, Psygmorchis, Quekettia, Rusbyella, Sanderella, Saundersia, Sigmatostalix, Solenidium, Symphyglossum, Trizeuxis.* (2) *Lophiaris, Trichocentrum.* (3) *Antillanorchis, Comparettia, Diadenium, Ionopsis, Neokoehleria, Oncidium* section *Oncidium* (the Variegata group), *Plectrophora, Polyotidium, Pterostemma, Rodriguezia, Rodrigueziopsis, Scelochilus.* (4) *Cischweinfia, Cypholoron, Helcia, Macradenia, Notylia, Oliveriana, Otoglossum, Psychopsis, Rossioglossum, Systeloglossum, Trichopilia, Warmingia.* (5) *Lockhartia.*

INTERGENERIC HYBRIDS   Three natural intergeneric hybrids are known in this group, all from Brazil. These are between *Oncidium* and *Ornitho-*

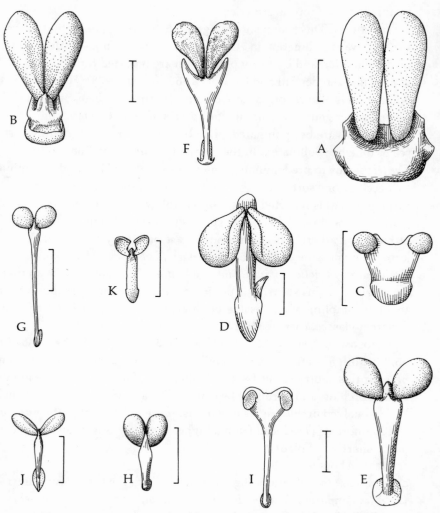

Figure 10.10 Representative pollinaria of the subtribe Oncidiinae. *(A) Oncidium (Lophiaris) cavendishianum. (B) Brassia arcuigera. (C) Systeloglossum costaricense. (D) Odontoglossum maculatum. (E) Comparettia macroplectron. (F) Trichopilia turialbae. (G) Oncidium cheirophorum. (H) Oncidium ansiferum. (I) Notylia bicolor. (J) Sigmatostalix guatemalense. (K) Hybochilus inconspicuus.* Scale 1 mm.

phora, between *Oncidium* and *Miltonia*, and between *Aspasia* and *Miltonia*. There are also a number of natural interspecific hybrids in *Miltonia*, and some of these would become intergeneric if *Miltonia* were split into two or more genera, as some authors have proposed. Such a division would be based on lip-shape only, and the natural hybrids

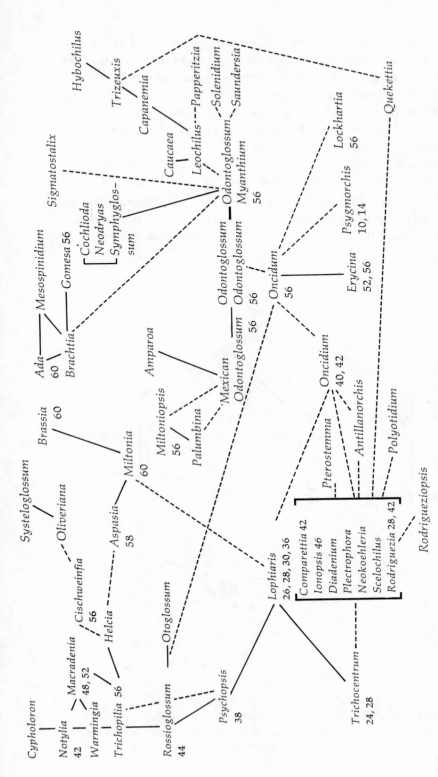

Figure 10.11 Tentative diagram of generic relationships in the subtribe Oncidiinae. Close and definite relationships (probably congeneric) are indicated by a heavy line. Solid lines indicate definite and rather close relationships, while dashed lines indicate relationships that are either more distant or less certain.

suggest that it would be no improvement on the present classification. Artificial intergeneric hybrids are known in a number of combinations (see figs. 10.12 and 10.13). Crosses between genera with different chromosome numbers are often harder to make, and the offspring more sterile, than in crosses involving more closely related genera, but it is likely that all the genera of this subtribe can be linked up directly or indirectly by crossability.

DISCUSSION   This group is very diverse in nearly all features, yet it is tied together by interfertility, and none of the proposed subdivisions is really consistent when carefully studied. N. H. Williams and I have been interested in this group for years, and our scheme of relationships is shown in figure 10.11. In the lists of genera, I have indicated some of

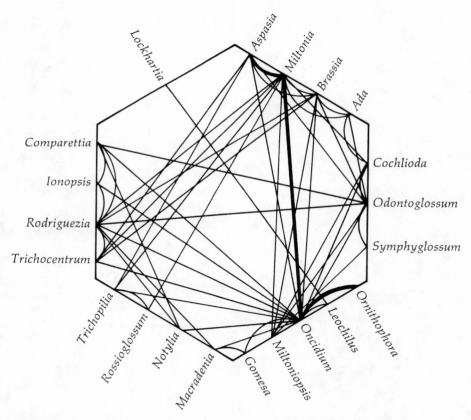

Figure 10.12   Intergeneric hybrids known in the subtribe Oncidiinae. Naturally occurring hybrids are indicated by heavy lines. In this diagram *Oncidium* has been treated in the broad, traditional sense. Some of the elements commonly included in *Oncidium* are treated separately in figure 10.13.

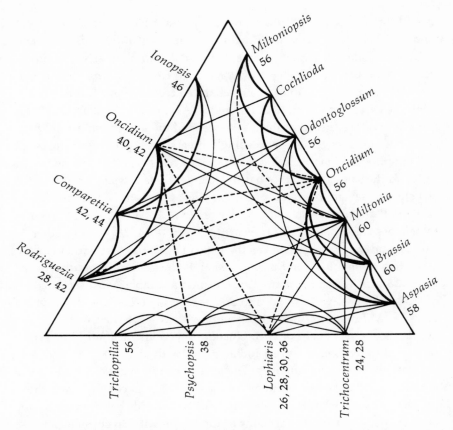

Figure 10.13 Intergeneric hybrids involving *Oncidium* and some allied genera. In this diagram, *Lophiaris*, *Psychopsis*, the West Indian *variegatum* complex (*Oncidium* section *Oncidium*) and the bulk of *Oncidium* species with 56 chromosomes are each treated separately. Heavy lines indicate combinations, some plants of which have proven to be fertile; dotted lines indicate combinations that are, as far as known, sterile; thin, continuous lines indicate hybrids for which I have no information on fertility. Diploid chromosome numbers are included, when known.

the more distinctive complexes, but these merge into each other too much to permit the delineation of distinct subtribes. *Lockhartia* is very unusual in its inflorescence and in the structure of its pollinaria, but a recent chromosome count indicates 56 chromosomes, and it has been crossed successfully with *Leochilus*, thus confirming its placement in this subtribe.

The Oncidiinae offer several examples of artificial classification. Until recently, *Oncidium, Odontoglossum,* and *Miltonia* were all characterized primarily by flower shape and the angle between the lip and column, and all three were quite artificial. Recent workers have separated the Andean "Miltonias" from the Brazilian group, as *Miltoniopsis,* and have transferred other discordant elements to *Oncidium.* Similarly, *Rossioglossum* and *Otoglossum* have been removed from *Odontoglossum.* The *Osmoglossum* group is once again separated from *Odontoglossum* but might be better combined with *Palumbia* (Ayensu and Williams, 1972). *Oncidium* offers special problems. As defined by flower shape and lip-column angle, this genus is exceedingly diverse in vegetative features and chromosome numbers. Dodson (1962b) has suggested that adaptation to the same pollinators has independently led to similar flower structure in several related groups, and this appears to be the case. The sections *Plurituberculata* and *Cebolletae* (2n = 28, 30, 32, 34, 36) seem to be more closely allied to *Trichocentrum* (2n = 24, 28) than to other genera, and may be separated as the genus *Lophiaris.* Also rather closely allied to *Lophiaris,* and to *Rossioglossum,* is the section *Glanduligera* (2n = 38), for which the name *Psychopsis* is available. Most sections of "*Oncidium*" have 2n = 56, and are more easily crossed with *Odontoglossum, Miltoniopsis,* and other genera with 2n = 56 than with *Lophiaris* or *Psychopsis,* or with *Oncidium* section *Oncidium.* The type section of *Oncidium* is a West Indian group with 2n = 40 or 42 and is vegetatively similar to *Ionopsis.* The species of this section cross easily with *Ionopsis* and other members of the *Comparettia* complex, and the hybrids are somewhat fertile (see fig. 10.13). Hybrids between *Oncidium* section *Oncidium* and the sections of *Oncidium* with 56 chromosomes, however, are quite sterile (Moir, 1978b). It seems clear that this West Indian complex should be removed from the main body of "*Oncidium.*" Unfortunately, the name *Oncidium* is fixed to the West Indian *O. variegatum,* so that this cannot be done without renaming the bulk of the species with 56 chromosomes.

REFERENCES   Charanasri, Kamemoto, and Takeshita, 1973 (chromosome numbers); Dod, 1976 (pollination of *Oncidium*); Garay and Stacy, 1974 (synopsis of *Oncidium*); Kränzlin, 1922 (revision of some genera); Williams, 1970 (pollinaria); Williams and Dressler, in prep. (classification).

Keys to Subfamilies, Tribes, and Subtribes

11

The keys given here will serve as a summary of the classification that has been presented. Let us, however, be very honest about it. Not all orchids will be unequivocally identified by these keys. For purposes of identification, it is better to use a quite "artificial" key that uses features like terminal inflorescence versus lateral inflorescence, or plicate leaves versus conduplicate leaves. In such a key, many very natural groups will have to be keyed out repeatedly, but if well made, the key will work. No such key has been made on a world-wide basis, and it would probably be nearly as long as the book. For purposes of identification, one should seek a key written for his own area, or for a nearby area. If the key was based on Schlechter's key (or on this one), it will work only for a certain percentage of the orchid family. Any key that attempts to key out subfamilies, tribes, and subtribes with simple, one-choice features cannot possibly work for all orchids.

To identify a plant by these keys, start with the first pair of alternatives, read both carefully before making a choice, and then go to the number indicated; this procedure is then repeated until one reaches the name, of the subfamily, tribe, or subtribe.

Key to Subfamilies and Anomalous Tribes

1. Flowers normally with two or three anthers each: 2
   Flowers normally with only one fertile anther each: 3
2. Lip deeply saccate; two fertile anthers; staminodia shieldlike:
   CYPRIPEDIOIDEAE
   Lip only slightly larger than petals, not deeply saccate; two or three
   fertile anthers; staminode absent or fingerlike: APOSTASIOIDEAE
3. Pollen generally soft and mealy, sectile or not; leaves usually spiral,
   convolute, not jointed basally: 4

Pollinia generally hard and waxy, two to eight; leaves distichous, usually
jointed basally: 8

4. Anther erect in early bud and bending down over apex of column to
become more or less operculate on apex of column: EPIDENDROIDEAE
(KEY III)

Anther remaining erect (or bending back), not short and operculate on
apex of column: 5

5. Anther more or less embedded in column; saprophytes with sectile
pollinia of many slender, ellipsoid massulae; roots ellipsoid
tropical America): WULLSCHLAEGELIEAE (ANOMALOUS TRIBE)

Anther not surrounded by columnar tissue; without the above
combination of features: 6

6. Rostellum subequal to anther; viscidium at apex of anther and attached
to apex of pollinia or caudicles; plants never with root-stem
tuberoids: SPIRANTHOIDEAE (KEY I)

Rostellum usually shorter than anther; anther usually projecting beyond
rostellum, viscidium, if present, usually at base or middle of pollinia;
plants often with root-stem tuberoids: 7

7. Plants with clustered, fleshy roots, or with root-stem tuberoids; often
rosette plants; leaves spiral, rarely plicate: ORCHIDOIDEAE (KEY II)

Plants with nodular tuberoids, or with the roots scattered on a
horizontal rhizome; leaves scattered and subdistichous or plicate:
TRIPHOREAE (ANOMALOUS TRIBE)

8. Anther erect in early bud and bending down over apex of column to
become operculate on apex of column; two, four, six, or eight pollinia,
usually laterally flattened or clavate, with or without caudicles,
with or without viscidia, usually without a stipe: EPIDENDROIDEAE
(KEY III)

Anther usually operculate on apex of column, but not bending downward during development; two or four pollinia, usually dorso-
ventrally flattened if four, with reduced caudicles, always with
viscidium and usually with a stipe or stipes: VANDOIDEAE (KEY IV)

Key I: SPIRANTHOIDEAE

1. Roots usually scattered on rhizome; pollinia sectile: (Erythrodeae) 2
Roots usually clustered; pollinia not sectile: (Cranichideae) 3

2. Stem woody; leaves plicate; inflorescence terminal or lateral: TROPIDIINAE
Stem herbaceous; leaves conduplicate; inflorescence terminal:
GOODYERINAE

3. Flowers resupinate: 4
   Flowers not resupinate: 6
4. Staminodia broad, conspicuous, clasping the anther and reaching
   beyond the middle of the anther: 5
   Staminodia small, inconspicuous, clasping only the base of the anther:
   SPIRANTHINAE
5. Column bent sharply up and down again; flower without an externally
   visible spur (Africa): MANNIELLINAE
   Column straight, flower with a prominent, external spur (New
   Caledonia): PACHYPLECTRONINAE
6. Column very short; lip free from column, much larger than the sepals
   and petals, which are similar and attenuate (Australasia and tropical
   Asia): CRYPTOSTYLIDINAE
   Not with the above combination of features (Americas and New
   Caledonia): CRANICHIDINAE

Key II: ORCHIDOIDEAE

1. Without root-stem tuberoids; leaves plicate or conduplicate; anther
   generally oblong: (Neottieae) 2
   Usually with root-stem tuberoids; leaves conduplicate; anther often
   conical: 3
2. Leaves (if present) plicate, scattered; with or without a viscidium:
   LIMODORINAE
   Leaves (if present) conduplicate, subopposite; rostellum sensitive,
   extruding a drop of glue when touched: LISTERINAE
3. Viscidium usually double; pollinia sectile, usually with prominent
   caudicles; anther solidly united with column (widespread): 9
   Viscidium single or lacking; pollinia sectile or not, usually lacking
   caudicles; with a constriction between anther and column (Australia,
   South America): (Diurideae) 4
4. Usually lacking a viscidium; root-stem tuberoids usually lacking (South
   America and New Caledonia): CHLORAEINAE
   With or without a viscidium; root-stem tuberoids usually present
   (Australasia): 5
5. Pollinia sectile; leaves onion-like (cylindrical and hollow):
   PRASOPHYLLINAE
   Pollinia not sectile, leaves various: 6
6. Plant with a single cordate or lobed leaf: ACIANTHINAE
   Leaves various, but not solitary and cordate: 7

7. Column short, thick, usually with arm-like or brush-like wings that are parallel with the column and may project beyond the anther; leaves usually very narrow: DIURIDINAE

Column longer, often arched, usually more or less flattened and bearing divergent wings; leaves usually broader: 8

8. Column with retrorse wings; sepals forming a hood over column; lip motile: PTEROSTYLIDINAE

Not with the above combination of features: CALADENIINAE

9. Anther erect; lip with a single basal spur (or none): (Orchideae) 10

Anther bent back from column (reclinate) or recumbent, with the base uppermost, spurs various, double if formed by lip: (Diseae) 12

10. Anther cells together above and diverging below; petals clawed and fimbriate (Africa): HUTTONAEINAE

Without the above combination of features: 11

11. Stigmas convex or stalked: HABENARIINAE

Stigma(s) concave: ORCHIDINAE

12. Lip erect, united with the face of the column, usually bearing an appendage which overtops the anther; dorsal sepal and petals commonly forming a hood: CORYCIINAE

Lip free from column: 13

13. Flowers resupinate; spur, when present, formed by dorsal sepal; column relatively short; anther suberect or reclinate: DISINAE

Flowers not resupinate; spurs, when present, formed by lip; column usually elongate; anther recumbent: SATYRIINAE

Key III: EPIDENDROIDEAE

1. Pollinia mealy or sectile, two, four, or too soft to form definite pollinia, without caudicles; leaves, if present, nonarticulate: 2

Pollinia either mealy or hard, if soft, usually eight and with caudicles; leaves usually articulate: 9

2. Pollinia soft and mealy: (Vanilleae) 3

Pollinia sectile: 6

3. Leaves distinctly plicate; pollinia coherent: PALMORCHIDINAE

Leaves fleshy, net-veined or lacking; pollinia very soft: 4

4. Usually fleshy vines, shrubs or saprophytes; stigma often emergent; seed coat hard: Vanillinae

Plants herbaceous; stigma not emergent; seed coat thin and papery: 5

5. Saprophytic plants with distinct calyculus beneath perianth (tropical Asia): LECANORCHIDINAE

Plants photosynthetic or saprophytic; without a distinct calyculus (Americas and northern Asia): POGONIINAE

6. Leaves narrow, plicate: ARETHUSINAE

Leaves broad, more or less fan-shaped, or plants saprophytic: (Gastrodieae) 7

7. Plants with leaves, not saprophytic; sepals free: NERVILIINAE

Plants leafless saprophytes; sepals often united: 8

8. Flowers racemose or solitary, usually borne above the soil level: GASTRODIINAE

Flowers densely clustered, usually borne within the soil or at the surface (Australia): RHIZANTHELLINAE

9. Saprophytes with sectile pollinia and caudicles: EPIPOGIEAE

Without the above combination of features: 10

10. Pollinia two or four, quite naked, without caudicles, rarely with viscidia or stipes: 11

Pollinia two to eight, with distinct caudicles (though sometimes very small): 14

11. Flowers with a prominent column foot; leaves always conduplicate; pollinia not clavate: 12

Flowers with a short or no column foot; leaves conduplicate or plicate; pollinia often clavate: MALAXIDEAE

12. Pseudobulbs usually of several internodes; inflorescence usually upper axillary or terminal; pollinia without viscidia: DENDROBIINAE

Pseudobulbs of one internode; inflorescence lateral or from rhizome; occasionally with viscidia or stipes: 13

13. Usually without viscidia or stipes; rarely with one stipe; anther operculate, deciduous: BULBOPHYLLINAE

With two distinct stipes; anther firmly attached to column, dehiscing away from clinandrium: SUNIPIINAE

14. Pollinia usually eight and rather soft; plants usually cormous; usually with plicate leaves and lateral inflorescence: (Arethuseae) 15

Pollinia two to eight, usually rather hard; plants with pseudobulbs, corms or slender stems (if corms, then of one internode); leaves usually conduplicate; inflorescence usually terminal: 17

15. Plants with slender stems (tropical America): SOBRALIINAE

Plants usually with corms or thick, fleshy pseudobulbs: 16

16. Pollinia sectile; leaves nonarticulate: ARETHUSINAE

Pollinia not sectile; leaves usually articulate: BLETIINAE

17. Pollinia four(?), soft and mealy; plants with elongate, leafy pseudobulbs and terminal inflorescence (Asia): THUNIINAE

Not with the above combination of features: 18

18. Pollinia superposed, with cylindrical, translucent caudicles or with well-developed stipe: 19

Pollinia superposed or not, with mealy caudicles, never with stipe: 20

19. Pollinia with cylindrical, translucent caudicles: CRYPTARRHENEAE

Pollinia with a definite, broad stipe: CALYPSOEAE

20. Pollinia four, superposed or ovoid, plants usually with pseudobulbs of one internode, with terminal inflorescence (Asia): (Coelogyneae) 21

Pollinia various, but not superposed; plants usually with pseudobulbs of several internodes or without pseudobulbs: (Epidendreae) 22

21. Plants without pseudobulbs: ADRORHIZINAE

Plants with pseudobulbs: COELOGYNINAE

22. Flowers with a joint between ovary and pedicel, pedicel persisting; without pseudobulbs (American): PLEUROTHALLIDINAE

Flowers never with a joint between ovary and pedicel: 23

23. American plants; inflorescence usually terminal: 24

Old World plants; inflorescence often lateral: 25

24. Pollinia eight, long-clavate; leaves solitary, fleshy: MEIRACYLLIINAE

Pollinia two to eight, usually laterally flattened or ovoid; habit various: LAELIINAE

25. Pollinia four, laterally flattened; inflorescence terminal: GLOMERINAE

Pollinia eight, or distinctly clavate; inflorescence usually lateral: 26

26. Pollinia two to eight, long-clavate; stems usually slender: PODOCHILINAE

Pollinia eight, ovoid, laterally flattened, or short-clavate: 27

27. Pollinia ovoid, attached to a common caudicle, this often very long; flowers tiny: THELASIINAE

Pollinia laterally flattened or short-clavate, not attached to a common caudicle; flowers usually larger: ERIINAE

Key IV: VANDOIDEAE

1. Plants always monopodial: (Vandeae) 2

Plants usually sympodial: 4

2. Lip saccate, but only rarely with a deep spur; pollinia four or two (mainly Asiatic): SARCANTHINAE

Lip usually deeply spurred; pollinia two (mainly African and American): 3

3. Rostellum deeply notched: ANGRAECINAE

Rostellum beaked: AERANGIDINAE

4. Pollinia four: 5

Pollinia two (usually cleft): 13

5. Plants with pseudobulbs of several internodes or elongate stems; inflorescence usually terminal (mainly African): POLYSTACHYEAE
   Plants various; inflorescence usually lateral (mainly American): (Maxillarieae) 6
6. Plants generally cormous, terrestrial; lip thin, sometimes spurred (mostly northern): CORALLORHIZINAE
   Plants various, usually epiphytic or without pseudobulbs; lip usually with a prominent callus, rarely spurred: 7
7. Plants small, monopodial, the one-flowered inflorescence arising opposite the leaf axil; column generally with a "ligule" beneath the stigma: DICHAEINAE
   Not with the above combination of features: 8
8. Pollinia flattened and superposed; generally larger plants: 9
   Pollinia usually ovoid or clavate, not markedly superposed; small plants: 12
9. Stipes long and narrow; leaves clearly plicate: LYCASTINAE
   Stipes short and wide, or lacking; leaves either plicate or conduplicate: 10
10. Leaves conduplicate, usually leathery or fleshy; viscidium wide, usually semilunate; inflorescence always one-flowered: MAXILLARIINAE
    Leaves plicate or conduplicate, usually thin; viscidium flattened, usually longer than wide; inflorescence various: 11
11. Callus usually prominent and with longitudinal ridges or keels; with or without pseudobulbs, may have pseudobulbs of several internodes: ZYGOPETALINAE
    Callus usually low, smooth or without keels; always with pseudobulbs of a single internode: BIFRENARIINAE
12. Column bristly; viscidium hooked: TELIPOGONINAE
    Column not bristly; viscidium not hooked; stigma usually at base of column; anther often long-beaked: ORNITHOCEPHALINAE
13. Plants generally with corms or pseudobulbs of several internodes; leaves usually plicate: 14
    Plants usually with pseudobulbs of one internode, or without pseudobulbs; leaves plicate or conduplicate: 16
14. Pollinia laterally flattened: ACRIOPSIDINAE
    Pollinia thick or dorsoventrally flattened: 15
15. Plants with sensitive rostellum, which discharges the viscidia when triggered; flowers unisexual or bisexual: CATASETINAE
    Rostellum not sensitive; flowers bisexual: CYRTOPODIINAE
16. Leaves plicate: STANHOPEINAE
    Leaves conduplicate: 17

17. Old World plants: 18
    American plants: 19
18. Lip and column highly united; column sigmoid (Asia): THECOSTELINAE
    Lip free from column; column short, with a long foot (Africa):
        GENYORCHIDINAE
19. Plants monopodial, often with two cylindrical, translucent caudicles
        longer than the stipe: PACHYPHYLLINAE
    Not with the above combination of features: ONCIDIINAE

What We Need to Know about Orchids

12

Although the preceding chapters have brought together our knowledge of some aspects of the orchids, they also clearly show that there are still gaping holes in our information about these plants. For example, I have had to write "not known" for the chromosome numbers of many groups, and for other large groups the counts that have been made are too few to be meaningful. We do not need more chromosome counts of the *Cattleya* alliance or of most Sarcanthinae; they are monotonously the same, anyway. We do need chromosome counts of many other groups, and especially of those with smaller flowers, and of the southern hemisphere terrestrials.

We still know too little about the details of flower structure. For instance, we have no detailed information of the structure of the stipes that occur in spiranthoid or epidendroid groups. It is clear that anther development and structure is critical in orchid classification, yet the only information we have on the subject is sixty years old. This, it seems to me, is one of the most urgent needs in the study of orchid morphology. I have tried to be careful and detailed in my observations of pollinaria, but they are still superficial and crude compared to what we should have. In some plant groups biochemical data have become very useful in classification. To date, the biochemical data on the orchids are mostly very spotty. We do have some work on orchid alkaloids, but, so far, it cannot be related to orchid classification.

In most of these areas, the most productive approach would be to concentrate on one or a few tribes or subtribes and try to get good coverage of the group under study. The many intergeneric hybrids created by orchidists stand out as a notable research opportunity for botanists, yet for the most part no one takes advantage of them, and we know little about the fertility of these hybrids. Some of these subjects are so technical as to be of interest mainly to professional botanists, or to others with access to microscopes and laboratory equipment. However, there is much that the amateur can do to advance our knowledge of orchids, especially amateurs who live in the tropics. One does not need to be able to pronounce *warscewiczii*, or have an advanced degree in botany, to make a real contribution to our knowledge of orchids. To be quite honest, much excellent botanical work has been done by people without degrees in botany, and some abominable work has been done by people with degrees.

Taxonomic Study

At the present time, most botanists concerned with orchids are working on "floristic" projects; that is, they are trying to write orchid floras for limited geographic regions. Such floras are much needed, but they are not enough. The floristic worker must focus on a particular area, and one cannot be sure whether the different names used in different regions really refer to different species or not. Indeed, the same names may be applied to different species in different areas.

To offset these problems, another type of taxonomic work is needed, in which one concentrates on a genus, a section of a genus, or some other taxonomic group and studies that group wherever it occurs. This kind of project gives a pretty clear idea of how many species there are in the group, where they occur, and what their correct names are. To do this sort of work properly, one needs to get (by loan) all the available museum specimens of the group under study; sketches and photographs are not an adequate substitute for the real specimens. (When critical specimens have been lost or destroyed, of course, we have to settle for whatever we can find.) After one has studied all the material, including living material and flowers preserved in alcohol, and worked out the species, one then determines what names to apply to the species. For this, the "type specimens" are critical. These are the specimens on which new species have been based. "Type specimens" are not necessarily "typical," but as the specimens on which new names were based, they are of historical value and serve as landmarks to determine the application of the names. The first specimen to be collected and named may be a very unusual one, but this does not affect its status as a type specimen.

Collecting Specimens for Taxonomic Study

Anyone who goes out and collects orchids in the tropics can make an important contribution by preparing museum specimens of the plants that he collects. While floristic work is going on in a number of areas, orchids are generally not well sampled. General plant collectors do not pay enough attention to them, and many orchid hobbyists hate to cut off a flower or a leaf to make a specimen. Thus, a great amount of interesting material is collected and cultivated, but because no record is ever made of it, this material contributes nothing to our general knowledge of orchids. In most cases one can make a good specimen without damaging the plant, and even a few pressed flowers with a sketch of the plant can be very useful. In collecting orchids for study, the single most important rule is to keep records. It doesn't matter if your plant never has a name tag on it, but it should be tagged immediately as to where it was collected.

Pressed specimens are not very attractive, but they are relatively permanent (if moth crystals are added regularly). Live plants are wonderful, but they are mortal: they die, labels get lost, and eventually nearly all living collections are broken up and dispersed. Flowers preserved in alcohol or formalin-acetic alcohol (FAA) are very useful, but they dry out easily, and a pickled collection demands constant attention (or better vials than are usually available). Photographs are wonderful, but for scientific purposes they should always be accompanied by a few pressed flowers. It is amazing how often one needs to see the other side of the petal or column, and one simply cannot dissect a photograph. A good example of what I mean is provided by the photographs of *Kefersteinia deflexipetala* and *K. parvilabris* in the *Orchid Digest* (34:124, 1970). One photograph shows a young flower, and the other shows a much older flower from a different angle. They simply cannot be compared. Drawings are extremely useful, often showing details that are lost in pressing, but even here a few pressed flowers should accompany the drawing.

Hágsater (1978) has given detailed information on how to prepare pressed specimens. In general, the specimen, or parts of it, are pressed in newspaper between cardboards, and heated air is circulated through the bundle of cardboards until the specimens are dry. Always label the specimens as to where and when they were collected. Do not leave it until later. It is best, also, to take notes on flower color and any other feature that cannot be seen from the dried specimen. Specimens must be kept dry, and moth crystals or paradichlorobenzene should be kept with the specimens, especially in the tropics. One may not be able to identify the pressed specimen at once, and there may be no taxonomist working in the area who has time to work with the material, but if they are deposited in a good herbarium or museum, they will last and will always be available for study.

Field Study

There are many aspects of orchid biology that are best studied in the field. We always need more information on what agents are pollinating the flowers in nature. If you are lucky enough to observe pollination, please press a few flowers and preserve a few insects, if possible. If there is ever any question of their identity, it can be rechecked. We need to know much more about fruit set, population structure, and distribution within and between vegetation types under natural conditions, especially in tropical areas. Mitidieri's interesting study (1956) of *Oncidium flexuosum* is a fine example of what can be done. Recently we had a very nice contribution by Braga (1978). We still need more information on germination in nature. For the more technically minded, we also need to know much more about mycorrhizae in nature.

For all these studies, the prime commandment is "Publish!" Share your knowledge and make it available to others. We have seen, within the last decade, the death of one who had a really enormous knowledge of the orchids of his adopted country. Unhappily, most of his knowledge died with him. I know of another case of one who has collected very extensively and who would like to be recognized for his contributions to science. When I suggested that he press specimens, he said "No! Pressing an orchid is like pressing a watermelon." If he had pressed a few specimens of every micro-orchid that he has seen in flower, he would, by now, have made a tremendous contribution to our knowledge of orchids. Instead, he sent thousands of specimens to botanical gardens, but many were dead and rotting on arrival, including ones that had been collected in flower. Most of the rest have proven their mortality without contributing very much to the botany of the orchids.

On the credit side, we may note the great contributions that Eric Östlund has made to our knowledge of Mexican orchids. He kept records, he made detailed sketches and notes, and he documented his material with pressed specimens. This material formed the basis for L. O. Williams' "Orchidaceae of Mexico" (1951). More recently, Glenn Pollard has made a very real contribution to our knowledge of the orchids of Mexico, as recorded in a commemorative issue of *Orquídea* (Mex.; vol. 7, no. 1). Now we have a very active group of *orquidófilos* in Mexico, and some of them are turning out first-rate botanical studies. Similarly, the contribution of "Stalky" and Nora Dunsterville to our knowledge of Venezuelan orchids has rarely been equalled, even by professional botanists. Finally, let me cite the case of the Sociedad Colombiana de Orquideología, in Medellín. Many of the members find more interest in the never-ending variety of the native species than in

imported hybrids. They have organized themselves, so that each one
concentrates on a genus or a subtribe and makes pressed specimens
of that group for their herbarium. This enthusiastic group has developed
within the last eight or ten years, and some of them are already making
valuable contributions to orchid botany. If we had a couple of such
groups in each tropical country, we would soon see an impressive
increase in our knowledge of this fascinating family.

REFERENCES

ANNOTATED LIST OF ORCHID FLORAS

GLOSSARY

INDEX

References

Ackerman, J. D. 1975. Reproductive biology of *Goodyera oblongifolia* (Orchidaceae). *Madroño* 20:191–198.

———. 1977. Biosystematics of the genus *Piperia* Rydb. (Orchidaceae), *Botanical Journal of the Linnaean Society* 75:245–270.

Ackerman, J. D., and M. R. Mesler. 1979. Pollination biology of *Listera cordata* (Orchidaceae). *American Journal of Botany* 66:820–824.

Ackerman, J. D., and N. H. Williams. In press a. Pollen morphology of the tribe Neottieae and its impact on the classification of the Orchidaceae. *Grana*.

———. In press b. Pollen morphology of the Chloraeinae (Orchidaceae: Diurideae) and related subtribes.

Adams, C. D. 1972. Orchidaceae. In *Flowering plants of Jamaica*. Mona, Jamaica: University of the West Indies.

Adams, R. M., and G. J. Goss. 1976. The reproductive biology of the epiphytic orchids of Florida. Pt. 3: *Epidendrum anceps* Jacquin, *American Orchid Society Bulletin* 45:488–492.

Allen, P. H. 1952. The swan orchids: a revision of the genus *Cycnoches. Orchid Journal* 1:173–184, 225–230, 249–254, 273–276, 297–303.

———. 1959a. Orchid hosts in the tropics. *American Orchid Society Bulletin* 28:243–244.

———. 1959b. *Mormodes lineatum*: a species in transition. *American Orchid Society Bulletin* 28:411–414.

Alphonso, A. G. 1976. The role of the botanic gardens in the conservation of orchid species. *Proceedings of the Eighth World Orchid Conference*, 323–325.

Ames, O. 1910. The genus *Habenaria* in North America. *Orchidaceae* 4:1–288.

———. 1922. A discussion of *Pogonia* and its allies in the northern United States. *Orchidaceae* 7:3–44.

Ames, O., and D. S. Correll. 1952–1953. Orchids of Guatemala. *Fieldiana: Botany* 26:1–727.

Arditti, J. 1966a. The production of fungal growth regulating compounds by orchids. *Orchid Digest* 30:88–90.

————. 1966b. Flower induction in orchids. *Orchid Review* 74:208–217.

Arp, G. K. 1977. The conservation of tropical orchids in the temperate zone greenhouse. *American Orchid Society Bulletin* 46:809–812.

Ayensu, E. S., and N. H. Williams. 1972. Leaf anatomy of *Palumbina* and *Odontoglossum* subgenus *Osmoglossum*. *American Orchid Society Bulletin* 41:687–696.

Baker, H. G. 1959. Reproductive methods as factors in speciation in flowering plants. *Cold Spring Harbor Symposia on Quantitative Biology* 24:177–191.

Baker, R. K. 1972. Foliar anatomy of the Laeliinae (Orchidaceae). Ph.D. diss., Washington University, St. Louis.

Baldwin, J. T., Jr., and B. M. Speese. 1957. Chromosomes of *Pogonia* and of its allies in the range of Gray's Manual. *American Journal of Botany* 44:651–653.

Ball, J. S. 1978. *Southern African epiphytic orchids.* Johannesburg: Conservation Press.

Balogh, P. 1979. Morfología del polen de la tribu Cranichideae Endlicher subtribu Spiranthinae Bentham (Orchidaceae). *Orquídea* (Mexico) 7:241–260.

Barthlott, W. 1976a. Struktur und Funktion des Velamen Radicum der Orchideen. *Proceedings of the Eighth World Orchid Conference*, 438–443.

————. 1976b. Morphologie der Samen von Orchideen im Hinblick auf Taxonomische und Funktionelle Aspekte. *Proceedings of the Eighth World Orchid Conference*, 444–453.

Barthlott, W., and B. Ziegler. In press. Über ausziehbare helicale Zellwandverdickungen als Haft-Apparat der Samenschalen von *Chiloschista lunifera* (Orchidaceae). *Berichte der Deutschen Botanischen Gesellschaft* 93.

Beer, J. G .1863. *Beiträge zur Morphologie und Biologie der Familie der Orchideen.* Vienna: Carl Gerold's Sohn.

Benzing, D. H. 1973. Mineral nutrition and related phenomena in Bromeliaceae and Orchidaceae. *Quarterly Review of Biology* 48:277–290.

————. 1978. The life history profile of *Tillandsia circinnata* (Bromeliaceae) and the rarity of extreme epiphytism among the Angiosperms. *Selbyana* 2:325–337.

Benzing, D. H., and J. Seemann. 1978. Nutritional piracy and host decline: a new perspective on the epiphyte–host relationship. *Selbyana* 2:133–148.

Blaxell, D. F. 1972. *Arthrochilus* F. Muell. and related genera (Orchidaceae) in Australasia. *Contributions from the New South Wales National Herbarium* 4:275–283.

Blossfeld, H. 1974. Der Tau als Lebensfaktor der Epiphyten. *Die Orchidee* 25:118–123.

Bosser, J. 1976. Le genre *Hederorkis* Thou. (Orchidaceae) aux Mascareignes et aux Seychelles. *Adansonia* 16:225–228.

Bouriquet, G. 1954. Le Vanillier et la Vanille dans le monde. In *Encyclopédie biologique*, vol. 46. Paris: Paul Lechavelier.

Braga, P. I. S. 1978. Aspectos biológicos das Orchidaceae de uma campina de Amazonia Central. *Acta Amazonica* 7 (supplement 2): 1–89.

Breddy, N. C., and W. H. Black. 1954. Orchid mycorrhiza and their application to seedling raising. *Orchid Journal* 3:57–61.

Bremer, K., and H.-E. Wanntorp. 1978. Phylogenetic systematics in botany. *Taxon* 27:317–329.

Brieger, F. G. 1977. On the maxillariinae (Orchidaceae) with sepaline spur. *Botanische Jahrbücher für Systematik, Pflanzengeschichte und Pflanzengeographie* 97:548–574.

———. *See also* Schlechter, 1970–.

Brown, M. J., J. F. Jenkin, N. P. Brothers, and G. R. Copson. 1978. *Corybas macranthus* (Hook.f.) Reichb.f. (Orchidaceae), a new record for Macquarie Island. *New Zealand Journal of Botany* 16:405–407.

Burgeff, H. 1932. *Saprophytismus und Symbiose: Studien an tropischen Orchideen*. Jena: Gustav Fischer.

———. 1936. *Die Samenkeimung der Orchideen*. Jena: Gustav Fischer.

———. 1959. Mycorrhiza of orchids. In *The orchids: a scientific survey*, ed. C. L. Withner. New York: Ronald Press.

Burns, W. T. 1961. *Sophronitis* hybridizing. *Orchid Digest* 25:5–15.

Butzin, F. 1971. Die Namen der supragenerischen Einheiten der Orchidaceae. *Willdenowia* 6:301–340.

———. 1974. Bestimmungschlüssel für die in Kultur genommenen Arten des Coelogyninae (Orchidaceae). *Willdenowia* 7:245–260.

Cady, L. 1962. Genus *Spiculaea* Lindl. in Australia. *Australian Plants* 1(10):11–13.

———. 1967. The genus *Cryptostylis* in Australia. *Australian Plants* 4:75–77, 91.

———. 1969. An illustrated check-list of the genus *Pterostylis*. *Australian Plants* 5:60–74.

Capesius, I., and W. Barthlott. 1975. Isotopen-Markierungen und Rasterelektronenmikroskopische Untersuchungen des Velamen radicum der Orchideen. *Zeitschrift für Pflanzenphysiologie* 75:436–448.

Charanasri, U., H. Kamemoto, and M. Takeshita. 1973. Chromosome numbers in the genus *Oncidium* and some allied genera. *American Orchid Society Bulletin* 42:518–524.

Chen, S. C. 1965. A primitive new orchid genus *Tangtsinia* and its meaning in phylogeny. *Acta Phytotaxonomica Sinica* 10:193–206.

Clifford, H. T., and W. K. Smith. 1969. Seed morphology and classification of Orchidaceae. *Phytomorphology* 19:133–139.

Correa, M. N. 1968. Rehabilitación del género *Geoblasta* Barb. Rodr. *Revista del Museo de la Plata, Sección Botánica* 11:69–74.

———. 1969. *Chloraea*, género sudamericano de Orchidaceae. *Darwiniana* 15:374–500.

Correll, D. S. 1950. *Native orchids of North America north of Mexico*. Waltham:

Chronica Botanica Company; rpt. Stanford: Stanford University Press, 1978.

Coster, C. 1926. Periodische Blüteerscheinungen in den Tropen. *Annales du Jardin Botanique de Buitenzorg* 35:125–162.

Crepet, W. L. 1979. Insect pollination: a paleontological perspective. *Bioscience* 29:102–108.

Cribb, F. J. 1978. A revision of *Stolzia* (Orchidaceae). *Kew Bulletin* 33:79–89.

Cruden, R. W. 1972. Pollinators in high-elevation ecosystems: relative effectiveness of birds and bees. *Science* 176:1439–1440.

Curtis, J. T. 1939. The relation of the specificity of orchid mycorrhizal fungi to the problem of symbiosis. *American Journal of Botany* 26:390–399.

Danesch, E., and O. Danesch. 1969–1972. *Orchideen Europas*. 3 vols. (Südeuropa, Mitteleuropa, and Ophryshybriden). Bern: Hallwag Verlag.

Daniels, G. S., and R. L. Rodríguez Caballero. 1972. Sobre la morfología del *Oncidium globuliferum*. *Orquideología* 7:79–84.

Darwin, C. 1888. *The Various contrivances by which orchids are fertilised by insects*, 2nd ed. London: John Murray.

Das, S. 1976. Flowering calendar of Coelogynes. *Orchid Review* 84:210–211.

Davis, P. H. 1978. The moving staircase: a discussion on taxonomic rank and affinity. *Notes from the Royal Botanic Garden Edinburgh* 36:325–340.

Davis, P. H., and V. H. Heywood. 1963. *Principals of angiosperm taxonomy*. Edinburgh: Oliver & Boyd.

Dockrill, A. W. 1969. *Australian indigenous orchids*. Vol. 1: *The epiphytes and tropical terrestrial species*. Sydney: Society for Growing Australian Plants.

Docters van Leeuwen, W. M. 1929. Kurze Mitteilung über Ameisen-Epiphyten aus Java. *Berichte der Deutschen Botanischen Gesellschaft* 47:90–99.

―――. 1936. Krakatau, 1833–1933. *Annales du Jardin Botanique de Buitenzorg* 46–47:1–506.

―――. 1937. The biology of *Epipogium roseum* (D. Don) Lindl. *Blumea*, supplement 1:57–65.

Dod, D. D. 1976. *Oncidium henekenii*—bee orchid pollinated by bee. *American Orchid Society Bulletin* 45:792–794.

Dodson, C. H. 1962a. Pollination and variation in the subtribe Catasetinae (Orchidaceae). *Annals of the Missouri Botanical Garden* 49:35–56.

―――. 1962b. The importance of pollination in the evolution of the orchids of tropical America. *American Orchid Society Bulletin* 31:525–534, 641–649, 731–735.

―――. 1966. Studies in orchid pollination—*Cypripedium*, *Phragmipedium* and allied genera. *American Orchid Society Bulletin* 35:125–128.

―――. 1975. *Dressleria* and *Clowesia*: a new genus and an old one revived in the Catasetinae (Orchidaceae). *Selbyana* 1:130–137.

Dodson, C. H., R. L. Dressler, H. G. Hills, R. M. Adams, and N. H. Williams. 1969. Biologically active compounds in orchid fragrances. *Science* 164:1234–1249.

Dressler, R. L. 1957. The vegetation about Laguna Ocotal. *Bulletin of the Museum of Comparative Zoology* 116:200–203.

——. 1960. The relationships of *Meiracyllium* (Orchidaceae). *Brittonia* 12:222–225.

——. 1961. The structure of the orchid flower. *Missouri Botanical Garden Bulletin* 49:60–69.

——. 1968a. Observations on orchids and euglossine bees in Panama and Costa Rica. *Revista de Biología Tropical* 15:143–183.

——. 1968b. Pollination by euglossine bees. *Evolution* 22:202–210.

——. 1971. Dark pollinia in hummingbird-pollinated orchids, or do hummingbirds suffer from strabismus? *American Naturalist* 105:80–83.

——. 1974. Classification of the orchid family. *Proceedings of the Seventh World Orchid Conference,* 259–279.

——. 1976a. How to study orchid pollination without any orchids. *Proceedings of the Eighth World Orchid Conference,* 534–537.

——. 1976b. The use of pollinia in orchid systematics. *First Symposium on Scientific Aspects of Orchids* (Detroit), 1–15.

——. 1979a. Una Enorme *Scaphyglottis* del Occidente de Panamá. *Orquídea* (Mexico) 7:227–234.

——. 1979b. The subfamilies of Orchidaceae. *Selbyana* 5:197–206.

——. 1980a. Orquídeas huérfanas. Pt. 1: *Wullschlaegelia. Orquídea* (Mexico) 7:277–282.

——. 1980b. Orquídeas huérfanas. Pt. 2: *Cryptarrhena. Orquídea* (Mexico) 7:283–288.

Dressler, R. L., and C. H. Dodson. 1960. Classification and phylogeny in the Orchidaceae. *Annals of the Missouri Botanical Garden* 47:25–68.

Dunsterville, G. C. K. 1961. How many orchids on a tree? *American Orchid Society Bulletin* 30:362–363.

Dunsterville, G. C. K., and E. Dunsterville. 1967. The flowering seasons of some Venezuelan orchids. *American Orchid Society Bulletin* 36:790–797.

Dunsterville, G. C. K., and L. A. Garay. 1959–1976. *Venezuelan orchids illustrated.* 6 vols. London: Andre Deutsch.

Duperrex, A. 1961. *Orchids of Europe.* London: Blandford Press.

Eberle, G. 1974. Nektarausscheidung im Sporn des Wanzenknabenkrautes (*Orchis coriophora* L.)? *Die Orchidee* 25:222–225.

Fawcett, W., and A. B. Rendle. 1910. Orchidaceae. In *Flora of Jamaica,* vol. 1. London: British Museum.

Foldats, E. 1969–1970. Orchidaceae. In *Flora de Venezuela,* vol. 15, ed. T. Lasser. Caracas: Instituto Botánico.

Fowlie, J. A. 1967. Observations on *Cattleya skinneri* and *C. deckeri. American Orchid Society Bulletin* 36:777–780.

——. 1970. *The genus* Lycaste. Pomona: Azul Quinta Press.

————. 1977. *The Brazilian bifoliate Cattleyas and their color varieties.* Pomona: Azul Quinta Press.

Frankie, G. W., P. A. Opler, and K. S. Bawa. 1976. Foraging behaviour of solitary bees: Implications for outcrossing of a Neotropical tree species. *Journal of Ecology* 64:1049–1057.

Frei, J. K. 1973a. Orchid ecology in a cloud forest in the mountains of Oaxaca, Mexico. *American Orchid Society Bulletin* 42:307–314.

————. 1973b. Effect of bark substrate on germination and early growth of *Encyclia tampensis* seeds. *American Orchid Society Bulletin* 42:701–708.

Fryxell, P. A. 1957. Mode of reproduction of higher plants. *Botanical Review* 23:135–233.

Garay, L. A. 1960. On the origin of the Orchidaceae. *Botanical Museum Leaflets* 19:57–95.

————. 1961. Notes on the genus *Epistephium. American Orchid Society Bulletin* 30:496–500.

————. 1964. Evolutionary significance of geographical distribution of orchids. *Proceedings of the Fourth World Orchid Conference,* 170–187.

————. 1969. El Complejo *Chondrorhyncha. Orquideología* 4:139–152.

————. 1972a. On the origin of the Orchidaceae. Pt. 2. *Journal of the Arnold Arboretum* 53:202–215.

————. 1972b. On the systematics of the monopodial orchids. Pt. 1. *Botanical Museum Leaflets* 23:149–212.

————. 1973. El Complejo *Zygopetalum. Orquideología* 8:15–34.

————. 1974a. Sinopsis del Género *Arpophyllum. Orquídea* (Mexico) 4:3–19.

————. 1974b. On the systematics of the monopodial orchids. Pt. 2. *Botanical Museum Leaflets* 23:369–376.

————. 1974c. *Acostaea* Schltr. y los Géneros del complejo *Pleurothallis. Orquideología* 9:103–124.

————. 1977. Systematics of the *Physurinae* (Orchidaceae) in the New World. *Bradea* 2:191–208.

————. 1979. Orchidaceae. Pt. 1. In *Flora of Ecuador,* vol. 9, ed. G. Harling and B. Sparre. Gothenburg: Department of Systematic Botany, University of Göteborg.

————. 1979. The genus *Phragmipedium. Orchid Digest* 43:133–148.

Garay, L. A., and J. E. Stacy. 1974. Synopsis of the genus *Oncidium. Bradea* 1:393–424.

Garay, L. A., and H. R. Sweet. 1974a. *Orchids of Southern Ryukyu Islands.* Cambridge: Botanical Museum.

————. 1974b. Orchidaceae. In *Flora of the Lesser Antilles,* ed. R. A. Howard. Jamaica Plain, Mass.: Arnold Arboretum.

Garay. L. A., and P. Taylor. 1976. The genus *Oecoclades* Lindl. *Botanical Museum Leaflets* 24:249–274.

Gentry, A. H. 1974. Flowering phenology and diversity in tropical Bignoniaceae. *Biotropica* 6:64–68.

Goss, G. J. 1977. The reproductive biology of the epiphytic orchids of Florida. Pt. 6: *Polystachya flavescens* (Lindley) J. J. Smith. *American Orchid Society Bulletin* 46:990–994.

Grant, V. 1949. Pollination systems as isolating mechanisms in angiosperms. *Evolution* 3:82–97.

———. 1958. The regulation of recombination in plants. *Cold Spring Harbor Symposia on Quantitative Biology* 23:337–363.

———. 1971. *Plant speciation.* New York: Columbia University Press.

Gregg, K. B. 1975. The effect of light intensity on sex expression in species of *Cycnoches* and *Catasetum* (Orchidaceae). *Selbyana* 1:101–113.

Grime, J. P. 1977. Evidence for the existence of three primary strategies in plants and its relevance to ecological and evolutionary theory. *American Naturalist* 111:1169–1194.

Gumprecht, R. 1975. Orchideen in Chile. *Die Orchidee* 26:127–132.

Hágsater, E. 1978. Preparación de especímenes para el herbario de orquídeas. *Orquídea* (Mexico) 6:369–394.

Hall, A. V. 1965. Studies of South African species of *Eulophia. Journal of South African Botany,* supplement 5:1–248.

Hallé, N. 1965. Deux Orchidées gabonaises présentées d'apres des sujets vivants: *Phaius mannii* Reichb. f. et *Manniella gustavi* Reichb. f. *Adansonia* 5:415–419.

———. 1977. Orchidacées. In *Flore de la Nouvelle Caledonie et dépendances,* ed. A. Aubreville and J.-F. Leroy. Paris: Museum National d'histoire Naturelle.

Hamer, F. 1974. *Las orquídeas del El Salvador.* 2 vols. San Salvador: Ministerio de Educación.

Harley, J. L. 1959. *The biology of mycorrhiza.* London: Leonard Hall.

Heinrich, B., and P. H. Raven. 1972. Energetics and pollination ecology. *Science* 1976:597–602.

Hennig, W. 1965. Phylogenetic systematics. *Annual Review of Entomology* 10:97–116.

Heusser, K. 1914. Die Entwicklung der generativen Organe von *Himantoglossum hircinum.* Thesis, Dresden.

Heywood, V. H., J. B. Harborne, and B. L. Turner. 1978. *The biology and chemistry of the Compositae.* 2 vols. London: Academic Press.

Hickey, L. J., and J. A. Doyle. 1977. Early Cretaceous fossil evidence for angiosperm evolution. *Botanical Review* 43:3–104.

Hills, H. G., N. H. Williams, and C. H. Dodson. 1968. Identification of some orchid fragrance components. *American Orchid Society Bulletin* 37:967–971.

Hirmer, M. 1920. Beiträge zur Organographie der Orchideenblüte. *Flora* 13:213–309.

Holman, R. T., and W. H. Heimermann. 1973. Identification of components of orchid fragrances by gas chromatography—mass spectrometry. *American Orchid Society Bulletin* 42:678–682.

Holttum, R. E. 1949. Gregarious flowering of the terrestrial orchid *Bromheadia finlaysonianum. The Garden's Bulletin* (Singapore) 12:295–302.

———. 1952a. Hybridization in Sarcanthinae. *Orchid Journal* 1:58–60.

———. 1952b. The subdivision of the genus *Dendrobium. Orchid Journal* 1:163–165.

———. 1955a. Growth habits of monocotyledons: variations on a theme. *Phytomorphology* 5:399–413.

———. 1955b. Notes on pollination of orchids of the genus *Plocoglottis. Malayan Nature Journal* 9:111–115.

———. 1957. *Orchids of Malaya,* 2nd ed. Singapore: Government Printing Office.

———. 1959. Evolutionary trends in the Sarcanthine orchids. *American Orchid Society Bulletin* 28:747–754 (also *Proceedings of the Second World Orchid Conference*).

———. 1960. The ecology of tropical epiphytic orchids. *Proceedings of the Third World Orchid Conference,* 196–203.

Hu, S. Y. 1977. *The genera of Orchidaceae in Hong Kong.* Hong Kong: Chinese University Press.

———. 1971–1975. The Orchidaceae of China. *Quarterly Journal of the Taiwan Museum* 24–28.

Huber, H. 1969. Die Samenmerkmale und Verwandschaftsverhältnisse der Liliifloren. *Mitteilungen der Botanischen Staatssamlung* (Munich) 7:219–538.

———. 1977. The treatment of the monocotyledons in an evolutionary system of classification. *Plant Systematics and Evolution,* supplement 1:284–298.

Hunt, T. E. 1953. Australia's subterranean orchids. *Orchid Journal* 2:303–305.

Ivri, Y., and A. Dafni. 1977. The pollination ecology of *Epipactis consimilis* Don (Orchidaceae) in Israel. *New Phytologist* 79:173–177.

Janzen, D. H. 1974. Epiphytic Myrmecophytes in Sarawak: mutualism through the feeding of plants by ants. *Biotropica* 6:237–259.

Johansson, D. R. 1974. Ecology of vascular epiphytes in West African rain forest. *Acta Phytogeographica Suecica* 59:1–129.

———. 1975. Ecology of epiphytic orchids in West African rain forests. *American Orchid Society Bulletin* 44:125–136.

———. 1977. Epiphytic orchids as parasites of their host trees. *American Orchid Society Bulletin* 46:703–707.

Jones, D. L. 1974. The pollination of *Acianthus caudatus* R. Br. *Victorian Naturalist* 91:272–274.

———. 1975. The pollination of *Microtis parviflora* R. Br. *Annals of Botany* 39:585–589.

Jones, D. L., and B. Gray. 1974. The pollination of *Calochilus holtzei* F. Muell. *American Orchid Society Bulletin* 43:604–606.

———. 1976a. The pollination of *Bulbophyllum longiflorum* Thouars. *American Orchid Society Bulletin* 45:15–17.

———. 1976b. The pollination of *Dendrobium lichenastrum* (F. Muell.) Krzl. *American Orchid Society Bulletin* 45:981–983.

———. 1977. The pollination of *Dendrobium ruppianum* A. D. Hawkes. *American Orchid Society Bulletin* 46:54–57.

Jones, K. 1967. The chromosomes of orchids. Pt. 2: Vandeae Lindl. *Kew Bulletin* 21:151–156.

———. 1974. Cytology and the study of orchids. In *The orchids: scientific studies,* ed. C. L. Withner. New York: John Wiley & Sons.

Jones, K., and M. G. Daker. 1968. The chromosomes of orchids. Pt. 3: Catasetinae Schltr. *Kew Bulletin* 22:421-427.

Kallunki, J. A. 1976. Population studies in *Goodyera* (Orchidaceae) with emphasis on the hybrid origin of *G. tesselata. Brittonia* 28:53–75.

Karasawa, K. 1979. Karyomorphological studies in *Paphiopedilum,* Orchidaceae. *Bulletin of the Hiroshima Botanical Garden* 2:1–149.

Kerr, A. D. 1971. A "trapdoor" *Eria. American Orchid Society Bulletin* 40:510–511.

Kleinfeldt, S. E. 1978. Ant-gardens: the interaction of *Codonanthe crassifolia* (Gesneriaceae) and *Crematogaster longispina* (Formicidae). *Ecology* 59:449–456.

Knudson, L. 1922. Non-symbiotic germination of orchid seeds. *Botanical Gazette* 73:1–25.

Kränzlin, F. 1910. Orchidaceae-Monandrae-Dendrobiinae. Pt. 1. *Das Pflanzenreich* 45:1–382.

———. 1911. Orchidaceae-Monandrae-Dendrobiinae. Pt. 2: Orchidaceae-Monandrae-Thelasiinae. *Das Pflanzenreich* 50:1–222.

———. 1922. Orchidaceae-Monandrae-Tribus Oncidieae-Odontoglosseae. Pt. 2. *Das Pflanzenreich* 80:1–344.

———. 1923. Orchidaceae-Monandrae-Pseudomonopodiales. *Das Pflanzenreich* 83:1–66.

———. 1925. Monographie der Gattungen *Masdevallia, Lothiana, Scaphosepalum, Cryptophoranthus* and *Pseudoctomeria. Repertorium Specierum Novarum Beiheft* 34:1–240.

———. 1926. Monographie der Gattung *Polystachya* Hook. *Repertorium Specierum Novarum Beiheft* 39:1–136.

———. 1928. Quelques orchidees nouvelles de la Nouvelle-Caledonie. *Notulae Systematicae* (Paris) 4:132–146.

Kullenberg, B. 1961. Studies on *Ophrys* pollination. *Zoologiska Bidrag Från Uppsala* 34:1–340.

Kullenberg, B., and G. Bergström. 1976. Hymenoptera Aculeata males as pollinators of *Ophrys* orchids. *Zoologica Scripta* 5:13–23.

Kumazawa, M. 1958. The sinker of *Platanthera* and *Perularia*: its morphology and development. *Phytomorphology* 8:137–145.

Landwehr, J. 1977. *Wilde Orchideeën van Europa*. 2 vols. Amsterdam: Verenigung tot Behoud van Natuurmonumenten in Nederland.

Larsen, K. 1969. Brief note on *Neuwiedia singapureana* in Thailand. *Natural History Bulletin of the Siam Society* 22:330–331.

Lavarack, P. S. 1971. *The taxonomic affinities of the Neottioideae*. 2 vols. Ph.D. diss., University of Queensland.

———. 1976. The taxonomic affinities of the Australian Neottioideae. *Taxon* 25:289–296.

León, Bro. 1946. Orquídeas. In *Flora de Cuba*, vol. 1, Contribuciones Ocasionales del Museo de Historia Natural del Colegio de La Salle.

Levin, D. A. 1978. The origin of isolating mechanisms in flowering plants. *Evolutionary Biology* 11:185–317.

Lin, T. P. 1975– . *Native orchids of Taiwan*. 2 vols. (of 4). Chiayi: Ji-Chyi Wang.

Lindley, J. 1826. *Orchidearum sceletos*. London.

Lock, J. M., and J. C. Profita. 1975. Pollination of *Eulophia cristata* (Sw.) Steud. (Orchidaceae) in southern Ghana. *Acta Botanica Neerlandica* 24:135–138.

Luer, C. A. 1972. *The native orchids of Florida*. New York: New York Botanical Garden.

———. 1975. *The native orchids of the United States and Canada, excluding Florida*. New York: New York Botanical Garden.

McKenna, M. C. 1975. Fossil mammals and early Eocene North Atlantic land continuity. *Annals of the Missouri Botanical Garden* 62:335–353.

McWilliams, E. L. 1970. Comparative rates of dark CO_2 uptake and acidification in the Bromeliaceae, Orchidaceae, and Euphorbiaceae. *Botanical Gazette* 131:285–290.

Madison, M. 1977. Vascular epiphytes: their systematic occurrence and salient features. *Selbyana* 2:1–13.

Maekawa, F. 1971. *The wild orchids of Japan in colour*. Tokyo: Seibundo Shinkosha.

Malguth, R. 1901. *Biologische Eigenthümlichkeiten der Früchte epiphytischer Orchideen*. Ph.D. diss., Breslau.

Mansfeld, R. 1932. Die Gattung *Catasetum* L. C. Rich. *Repertorium Specierum Novarum Regni Vegetabilis* 30:257–275, 31:99–125.

———. 1934. Zur Terminologie der Pollinienanhängsel der Orchideen. *Repertorium Specierum Novarum Regni Vegetabilis* 38:199–205.

———. 1937a. Über das System der Orchidaceae-Monandrae. *Notizblatt des Botanischen Gartens und Museums zu Berlin-Dahlem* 13:666–676.

————. 1937b. Ueber das System der Orchidaceae. *Blumea,* supplement 1:25–37.

————. 1954. Über die Verteilung der Markmale innerhalb der Orchidaceae-Monandrae. *Flora* 142:65–80.

Mayr, E. 1965. Numerical phenetics and taxonomic theory. *Systematic Zoology* 14:73–97.

————. 1969. *Principles of systematic zoology.* New York: McGraw-Hill.

————. 1974. Cladistic analysis or cladistic classification? *Zeitschrift für Zoologische und Systematische Evolutionsforschung* 12:94–128.

Millar, A. 1978. *Orchids of Papua New Guinea: an introduction.* Seattle: University of Washington Press.

Misas U., G., and O. J. Arango T. 1974. Introducción al conocimiento de una subtribu. *Orquideología* 9:47–60.

Mitidieri, J. 1956. Indice de sobrevivencia natural no *Oncidium flexuosum. Orquidea* 18:12–15.

Moir, W. W. G. 1974. Intergenerics in the Oncidiinae. *Orchid Review* 82:156–160.

————. 1975a. Intergenerics of the Lesser Laeliinae. *Florida Orchidist* 18:61–64.

————. 1975b. The wasted efforts in breeding. *Orchid Review* 83:298–302.

————. 1978a. Surprises in breeding orchids. *Orchid Review* 86:161–164.

————. 1978b. Breeding the *Oncidium* sect. *Oncidium* (erroneously the Equitant-Variegata Oncidiums). *Orchid Digest* 42:85–91.

Morris, B. 1970. *The epiphytic orchids of Malawi.* Blantyre: Society of Malawi.

Mulay, B. N., and B. D. Deshpande. 1959. Velamen in terrestrial monocots. Pt. 1: Ontogeny and morphology of velamen in the Liliaceae. *Journal of the Indian Botanical Society* 38:383–390.

Mulay, B. N., B. D. Deshpande, and H. B. Williams. 1958. Study of velamen in some epiphytic and terrestrial orchids. *Journal of the Indian Botanical Society* 37:123–127.

Mulay, B. N., and T. K. B. Panikar. 1956. Origin, development and structure of velamen in the roots of some species of terrestrial orchids. *Proceedings of the Rajasthan Academy of Science* 6:31–48.

Neales, T. F., and C. S. Hew. 1975. Two types of carbon fixation in tropical orchids. *Planta* (Berlin) 123:303–306.

Newton, G. D., and N. H. Williams. 1978. Pollen morphology of the Cypripedioideae and Apostasioideae (Orchidaceae). *Selbyana* 2:169–182.

Nicholls, W. H. 1969. *Orchids of Australia.* Melbourne: Nelson.

Nierenberg, L. 1972. The mechanism for the maintenance of species integrity in sympatrically occurring equitant Oncidiums in the Caribbean. *American Orchid Society Bulletin* 41:873–882.

Northen, R. T. 1952. Pollen-shooting mechanism of *Catasetum. American Orchid Society Bulletin* 21:859–862.

————. 1970. The mysterious movements of pollinaria. *Orchid Digest* 34:87–88.

————. 1971. The remarkable thrift of *Spiculaea ciliata*. *American Orchid Society Bulletin* 40:898–899.

————. 1972. *Pterostylis* and its sensitive gnat trap. *American Orchid Society Bulletin* 41:801–806.

Nuernbergk, E. L. 1963. On the carbon dioxide metabolism of orchids and its ecological aspect. *Proceedings of the Fourth World Orchid Conference*, 158–169.

Ogura, T. 1953. Anatomy and morphology of the subterranean organs in some Orchidaceae. *Journal of the Faculty of Science of the University of Tokyo: Botany* 6:135–157.

Ospina H., M. 1969. Los Antipolinizadores. *Orquideología* 4:23–27.

Ortiz Valdivieso, P. 1976. *Orquídeas de Colombia*. Bogotá: Colciencias.

Pabst, G. F. J. 1978. An illustrated key to the species of the genus *Mormodes* Lindl. (Orchidaceae). *Selbyana* 2:149–155.

Pabst, G. F. J. and F. Dungs. 1075–1977. *Orchidaceae Brasilienses*. 2 vols. Hildesheim: Kurt Schmersow.

Peisl, P., and J. Forster. 1975. Zur Bestäubungsbiologie des Knabenkrautes *Orchis coriophora* L. ssp. *fragrans*. *Die Orchidee* 26:172–173.

Peitz, E. 1972. Zusammenstellung aller bisher bekannte Bastarde der in Deutschland verbreiteten Orchideen. In *Probleme der Orchideengattung Orchis*, ed. K. Senghas and H. Sundermann. *Die Orchidee*, Sonderheft.

Perrier de la Bathie, H. 1939–1941. Orchidées. In *Flore de Madagascar*, ed. H. Humbert. 2 vols. Tananarive: Imprimerie Officielle.

Perry, D. R. 1978. Factors influencing arboreal epiphytic phytosociology in Central America. *Biotropica* 10:235–237.

Pfitzer, E. 1882. *Grundzüge einer Vergleichenden Morphologie der Orchideen*. Heidelberg: Carl Winter's Universitätsbuchhandlung.

————. 1884. Beobachtungen über Bau und Entwicklung der Orchideen. Pt. 9: Ueber Zwergartige Bulbophyllen mit Assimilationshöhlen im innern der Knollen. *Berichte der Deutschen Botanischen Gesellschaft* 2:472–480.

————. 1887. *Entwurf einer Natürlichen Anordnung der Orchideen*. Heidelberg: Carl Winter's Universitätsbuchhandlung.

————. 1903. Orchidaceae-Pleonandrae. *Das Pflanzenreich* 12:1–132.

Pfitzer, E., and F. Kränzlin. 1907. Orchidaceae-Monandrae-Coelogyninae. *Das Pflanzenreich* 32:1–169.

Piers, F. 1968. *Orchids of East Africa*, 2nd ed. Germany: J. Cramer.

van der Pijl, L., and C. H. Dodson. 1966. *Orchid flowers: their pollination and evolution*. Coral Gables: University of Miami Press.

Pollard, G. E. 1973. La Opinión de un Hombre. *Orquídea* (Mexico) 3:184–190.

Porsch, O. 1909. Die Honigersatzmittel der Orchideenblüte. *Botanische Wandtafeln*, plates 92, 111: 496–509.

Potůček, O. 1968. Intergenerische Hybriden der Gattung *Dactylorhiza*. In *Probleme der Orchideengattung Dactylorhiza*, ed. K. Senghas and H.

Sundermann. *Die Orchidee*, Sonderheft.

Pradhan, U. C. 1976. *Indian orchids: guide to identification and culture.* Vol. 1. Calcutta: Bharat Lithographing Company.

Pridgeon, A. M. 1978. Una Revisión de los Géneros *Coelia* y *Bothriochilus. Orquídea* (Mexico) 7:57–94.

Proctor, M., and P. Yeo. 1972. *The pollination of flowers.* New York: Taplinger Publishing Company.

Ramsey, C. T. 1950. The triggered rostellum of the genus *Listera. American Orchid Society Bulletin* 19:482–485.

Rao, V. S. 1969. The floral anatomy and relationship of the rare Apostasias. *Journal of the Indian Botanical Society* 48:374–385.

———. 1974. The relationships of the Apostasiaceae on the basis of floral anatomy. *Botanical Journal of the Linnaean Society* 68:319–327.

Rasmussen, F. N. 1977. The genus *Corymborkis* Thou. (Orchidaceae): taxonomic revision. *Botanisk Tidsskrift* 71:161–192.

Raven, P. H., and D. I. Axelrod. 1974. Angiosperm biogeography and past continental movements. *Annals of the Missouri Botanical Garden* 61:539–673.

Reichenbach, H. G. 1852. *De Pollinis Orchidearum Genesi ac Structura et de Orchideis in Artem ac Systema Redigendis.* Thesis, Leipzig.

———. 1884. Ueber das System der Orchideen. *Bulletin du Congres international de botanique et d'horticulture a St.-Petersbourg* 1884:39–58.

Rentoul, J. N. 1977. Australian bulbophyllums. *Orchid Review* 85:261–262.

Renz, J. 1948. Beiträge zur Kenntnis der süd- und zentralamerikanischen Orchideen. Pt. 1: Orchidaceae-Cranichidinae. *Candollea* 11:243–276.

———. 1978. Orchidaceae. In *Flora Iranica*, ed. K. H. Rechinger. Graz: Akademische Druck- und Verlagsanstalt.

Ridley, H. N. 1910. Symbiosis of ants and plants. *Annals of Botany* 24:457–483.

Riveros, M., and C. Ramírez. 1978. Fitocenosis Epífitas de la Asociación *Lapagerio-Aextoxiconetum* en el Fundo San Martín (Valdivia, Chile). *Acta Científica Venezolana* 29:163–169.

Rolfe, R. A. 1909–1912. The evolution of the Orchidaceae. *Orchid Review* 17–20.

Rosso, S. W. 1966. The vegetative anatomy of the Cypripedioideae (Orchidaceae). *Journal of the Linnaean Society: Botany* 59:309–341.

Rotherham, E. R. 1968. Pollination of *Spiculaea huntiana* (elbow orchid). *Victorian Naturalist* 85:7–8.

Rotor, G. B., Jr. 1952. Daylength and temperature in relation to growth and flowering of orchids. *Cornell University Agricultural Experimental Station Bulletin* 885:1–47.

Ruinen, J. 1953. Epiphytosis: a second view on epiphytism. *Annales Bogoriensis* 1:101–157.

Sanford, W. W. 1964. Sexual compatibility relationships in *Oncidium* and related genera. *American Orchid Society Bulletin* 33:1035–1048.

————. 1971. The flowering time of West African Orchids. *Botanical Journal of the Linnean Society* 65:163–181.

————. 1974. The ecology of orchids. In *The orchids: scientific studies,* ed. C. L. Withner. New York: John Wiley & Sons.

Santapau, H., and Z. Kapadia. 1966. *The orchids of Bombay.* Calcutta: Government of India Press.

Schelpe, E. 1966. *An introduction to the South African orchids.* London: Macdonald.

————. 1970. Fire-induced flowering among indigenous South African orchids. *South African Orchid Journal* 1(2):21–22.

————. 1971. The genus *Disa* and allied genera in South Africa. *Proceedings of the Sixth World Orchid Conference,* 157–159.

Schill, R. 1978. Palynologische Untersuchungen zur systematischen Stellung der Apostasiaceae. *Botanische Jahrbücher für Systematik, Pflanzengeschichte und Pflanzengeographie* 99:353–362.

Schill, R., and W. Pfeiffer. 1977. Untersuchungen an Orchideenpollinien unter besonderer Beruechsichtigung ihrer Feinskulpturen. *Pollen et Spores* 19:5–118.

Schlechter, R. 1911. Die Polychondreae (Neottiinae Pfitz.) und ihre systematische Einteilung. *Botanische Jahrbücher für Systematik, Pflanzengeschichte und Pflanzengeographie* 45:375–410.

————. 1920. Versuch einer systematischen Neuordnung der *Spiranthinae.* *Beiheft zum Botanisches Centralblatt* 37(2):317–454.

————. 1926. Das System der Orchidaceen. *Notizblatt des Botanischen Gartens und Museums zu Berlin-Dahlem* 9:563-591.

————. 1970–   . *Die Orchideen,* 3rd ed,. ed. F. G. Brieger, R. Maatsch, and K. Senghas. Berlin: Paul Parey.

Schmid, R., and M. J. Schmid. 1977. Fossil history of the Orchidaceae. In *Orchid biology: reviews and perspectives,* vol. 1, ed. J. Arditti. Ithaca: Comstock Publishing Associates.

Schultes, R. E. 1960. *Native orchids of Trinidad and Tobago.* Oxford: Pergamon Press.

Schweinfurth, C. 1958–1959. Orchids of Peru. *Fieldiana: Botany* 30:1–531.

————. 1959. Classification of orchids. In *The orchids: a scientific survey,* ed. C. L. Withner. New York: Ronald Press.

Schweinfurth, C., and D. S. Correll. 1940. The genus *Palmorchis. Botanical Museum Leaflets* 8:109–119.

Seidenfaden, G. 1968. The genus *Oberonia* in mainland Asia. *Dansk Botanisk Arkiv* 25(3):1–125.

————. 1969. Notes on the genus *Ione. Botanisk Tidsskrift* 64:205–238.

————. 1971. Notes on the genus *Luisia. Dansk Botanisk Arkiv* 27(4):1-101.

————. 1973. Notes on *Cirrhopetalum* Lindl. *Dansk Botanisk Arkiv* 29(1):1–260.

————. 1975a. Orchid genera in Thailand. Pt. 1: *Calanthe* R. Br. *Dansk Botanisk Arkiv* 29(2):1–50.

————. 1975b. Orchid genera in Thailand. Pt. 2: *Cleisostoma* B1. *Dansk*

Botanisk Arkiv 29(3):1–80.

———. 1975c. Orchid genera in Thailand. Pt. 3: *Coelogyne* Lindl. *Dansk Botanisk Arkiv* 29(4):1–94.

———. 1976. Orchid genera in Thailand. Pt. 4: *Liparis* L. C. Rich. *Dansk Botanisk Arkiv* 31(1):1–105.

———. 1977. Orchid genera in Thailand. Pt. 5: Orchidoideae. *Dansk Botanisk Arkiv* 31(3):1–149.

———. 1978a. Orchid genera in Thailand. Pt. 6: Neottioideae Lindl. *Dansk Botanisk Arkiv* 32(2):1–195.

———. 1978b. Orchid genera in Thailand. Pt. 7: *Oberonia* Lindl. & *Malaxis* Sol. ex Sw. *Dansk Botanisk Arkiv* 33(1):1–94.

Seidenfaden, G., and T. Smitinand. 1959–1965. *The orchids of Thailand: a preliminary list.* Bangkok: The Siam Society.

Senghas, K. 1973. Unterfamilie: Orchidoideae. In R. Schlechter, *Die Orchideen*, 3rd ed., vol. 1, ed. F. G. Brieger, R. Maatsch, and K. Senghas. Berlin: Paul Parey.

Senghas, K., and H. Sundermann. 1968. Probleme der Orchideengattung *Dactylorhiza. Die Orchidee*, Sonderheft.

———. 1970. Probleme der Orchideengattung *Epipactis. Die Orchidee*, Sonderheft.

———. 1972. Probleme der Orchideengattung *Orchis. Die Orchidee*, Sonderheft.

Sharman, B. C. 1939. The development of the sinker of *Orchid mascula* Linn. *Botanical Journal of the Linnean Society* 52:145–158.

Sheviak, C. J. 1974. An introduction to the ecology of the Illinois Orchidaceae. *Illinois State Museum Science Papers* 14:1–89.

Siebe, M. 1903. *Ueber den anatomischen Bau der Apostasiinae.* Ph.D. diss., Heidelberg.

Simpson, G. G. 1961. *Principles of animal taxonomy.* New York: Columbia University Press.

Smith, A. G., and J. C. Briden. 1977. *Mesozoic and Cenozoic paleocontinental maps.* Cambridge: Cambridge University Press.

Smith, G. R., and G. E. Snow. 1976. Pollination ecology of *Platanthera* (*Habenaria*) *ciliaris* and *P. blephariglottis* (Orchidaceae). *Botanical Gazette* 137:133–140.

Smith, J. J. 1925. Ephemeral orchids. *Annales du Jardin Botanique de Buitenzorg* 35:55–70.

Stebbins, G. L. 1970. Adaptive radiation of reproductive characteristics in angiosperms. Pt. 1: Pollination mechanisms. *Annual Reviews of Ecology and Systematics* 1:307–326.

———. 1974. *Flowering plants: evolution above the species level.* Cambridge: Belknap Press of Harvard University Press.

Stebbins, G. L., and L. Ferlan. 1956. Population variability, hybridization, and introgression in some species of *Ophrys. Evolution* 10:32–46.

Stewart, J. 1976. The Vandaceous group in Africa and Madagascar. *Proceedings of the Eighth World Orchid Conference*, 239–248.

Stewart, J., and B. Campbell. 1970. *Orchids of tropical Africa*. London: W. H. Allen.

Stojanow, N. 1916. Über die vegetative Fortpflanzung der Ophrydineen. *Flora* 109:1–39.

Stoutamire, W. P. 1963. Terrestrial orchid seedlings. *Australian Plants* 2:119–122.

———. 1964. Seeds and seedlings of native orchids. *Michigan Botanist* 3:107–119.

———. 1967. Flower biology of the Lady's-slippers (Orchidaceae: *Cypripedium*). *Michigan Botanist* 6:159–175.

———. 1971. Pollination in temperate American orchids. *Proceedings of the Sixth World Orchid Conference*, 233–243.

———. 1974a. Relationships of the purple-fringed orchids *Platanthera psycodes* and *P. grandiflora*. *Brittonia* 26:42–58.

———. 1974b. Terrestrial orchid seedlings. In *The orchids: scientific studies*, ed. C. L. Withner. New York: John Wiley & Sons.

———. 1975. Pseudocopulation in Australian orchids. *American Orchid Society Bulletin* 44:226–233.

———. 1978. Pollination of *Tipularia discolor*, an orchid with modified symmetry. *American Orchid Society Bulletin* 47:413–415.

Summerhayes, V. S. 1951. *Wild orchids of Britain*. London: Collins.

———. 1956. A revision of the genus *Brachycorythis*. *Kew Bulletin* 1956:221–264.

———. 1957. The genus *Eulophidium* Pfitz. *Bulletin du Jardin Botanique d'l'Etat* (Brussels) 27:391-403.

———. 1966. African orchids. *Kew Bulletin* 20:165–199.

———. 1968. Orchidaceae. Pt. 1. In *Flora of tropical East Africa*, ed. F. Milne-Redhead and R. M. Polhill. London: Crown Agents for Oversea Governments and Administrations.

Sundermann, H. 1964. Probleme der Orchideengattung *Ophrys*. *Die Orchidee*, Sonderheft.

———. 1975. *Europäische und mediterrane Orchideen*, 2nd ed. Hildesheim: Kurt Schmersow.

Swamy, B. G. L. 1948. Vascular anatomy of orchid flowers. *Botanical Museum Leaflets* 13:61–95.

Sweet, H. R. 1968–1969. A revision of the genus *Phalaenopsis*. *American Orchid Society Bulletin* 37:867–877, 1089–1104; 38:33–42, 225–239, 321–336, 505–519, 681–694, 888–901.

———. 1969. Orquídeas Andinas poco conocidas. Pt. 1: *Monophyllorchis*. *Orquideología* 4:179–181.

———. 1972. The genus *Porroglossum*. *American Orchid Society Bulletin* 41:513–524.

Tan, K. W. 1969. The systematic status of the genus *Bletilla* (Orchidaceae). *Brittonia* 21:202–214.

————. 1975–1976. Taxonomy of *Arachnis, Armodorum, Esmeralda* and *Dimorphorchis* (Orchidaceae). *Selbyana* 1:1–15, 365–373.

Tanaka, R. 1976. Cytological studies on the wide-crossing in Orchidaceae (in Japanese). *Recent Advances in Breedings* 12:95–112.

Tanaka, R., and H. Kamemoto. 1961. Meiotic chromosome behavior in some intergeneric hybrids of the *Vanda* alliance. *American Journal of Botany* 48:573–582.

Teoh, S. B., and S. N. Lim. 1978. Cytological studies in Malayan members of the *Phaius* tribe (Orchidaceae). Pt. 1: Somatic chromosomes. *Malaysian Journal of Science* 5(A)(978):1–11.

Thien, L. B. 1973. Isolating mechanisms in the genus *Calopogon. American Orchid Society Bulletin* 42:794–797.

Thien, L. B., and B. G. Marcks. 1972. The floral biology of *Arethusa bulbosa, Calopogon tuberosus* and *Pogonia ophioglossoides* (Orchidaceae). *Canadian Journal of Botany* 50:2319–2325.

Thien, L. B., and F. Utech. 1970. The mode of pollination in *Habenaria obtusata* (Orchidaceae). *American Journal of Botany* 57:1031–1035.

Tuyama, T. 1967. On *Epipogium roseum* (D. Don) Lindl. in Japan and its adjacent regions, with remarks on other species of the genus. *Journal of Japanese Botany* 42:295–311.

Ule, E. 1904. Ameisengärten im Amazonasgebiet. *Engler's Botanische Jahrbücher* 30(Beiblatt):45–52.

Vareschi, V. 1976. Orchideen und ihre ökologischen Nischen in den Tropen. *Proceedings of the Eighth World Orchid Conference*, 516–527.

Verdcourt, B. 1968. African Orchids. Pt. 31: New taxa of *Disperis* from East and Central Africa. *Kew Bulletin*, 22:93–99.

Vermeulen, P. 1947. Studies on Dactylorchids. Ph.D. diss., University of Amsterdam.

————. 1959. The different structure ofthe Rostellum in Ophrydeae and Neottieae. *Acta Botanica Neerlandica* 8:338–355.

————. 1965. The place of *Epipogium* in the system of the Orchidales. *Acta Botanica Neerlandica* 14:230–241.

————. 1966. The system of the orchidales. *Acta Botanica Neerlandica* 15:224–253.

Veyret, Y. 1965. Embroyogénie comparée et blastogénie chez les Orchidaceae-Monandrae. *Memoires Office de la Recherche Scientifique et Technique Outre-Mer* 12:1–106.

de Vogel, E. F. 1969. Monograph of the tribe Apostasieae (Orchidaceae). *Blumea* 17:313–350.

Vogel, S. 1954. Blütenbiologische Typen als Elemente der Sippengliederung dargestellt anhand der Flora Südafrikas. *Botanische Studien* 1:1–338. Jena: Gustav Fischer Verlag.

————. 1959. Organographie der Blüten Kapländischer Ophrydeen. *Abhandlungen der Akademie der Wissenschaften und der Literatur, Mathematisch-Naturwissenschaftliche Klasse* (Mainz) 1959:268–532.

————. 1962. Duftdrüsen im Dienste der Bestäubung. *Abhandlungen der Akademie des Wissenschaften und der Literatur, Mathematisch-Naturwissenschaftliche Klasse* (Mainz) 1962:602–763.

————. 1972. Pollination von *Orchis papilionacea* L. in den Schwarmbahnen von *Eucera tuberculata* F. *Die Orchidee*, Sonderheft.

————. 1974. Ölblumen und ölsammelnde Bienen. *Tropische und subtropische Pflanzenwelt* 7:1–267.

————. 1978. Pilzmückenblumen also Pilzmimeten. *Flora* 167:329–398.

Vöth, W. 1975. *Trielis villosa* var. *rubra*, Bestäuber von *Orchis coriophora*. *Die Orchidee* 26:170–172.

Warcup, J. H. 1975. Factors affecting symbiotic germination of orchid seed. In *Endomycorrhiza*, ed. F. E. Sanders, B. Mosse, and P. B. Tinker. London: Academic Press.

Waters, V. H., and C. C. Waters. 1973. *A survey of the slipper orchids*. Shelby: Carolina Press.

Weber, N. A. 1943. Parabiosis in neotropical "ant gardens." *Ecology* 24:400–404.

Went, F. W. 1940. Soziologie der Epiphyten eines tropischen Urwaldes. *Annales du Jardin Botanique de Buitenzorg* 50:1–98.

Wheeler, W. M. 1921. A new case of parabiosis and the "ant gardens" of British Guiana. *Ecology* 2:89–103.

Whittaker, R. H. 1969. New concepts of kingdoms of organisms. *Science* 163:150–160.

Williams, L. O. 1951. The Orchidaceae of Mexico. *Ceiba* 2:1–321.

————. 1956. An enumeration of the Orchidaceae of Central America, British Honduras and Panama. *Ceiba* 5:1–256.

Williams, L. O., and P. A. Allen. 1946–1949. Orchidaceae. In *Flora of Panama*, ed. R. E. Woodson and R. W. Schery. *Annals of the Missouri Botanical Garden* 33(1,4), 36(1,2).

Williams, N. H. 1970. Some observations on pollinaria in the Oncidiinae. *American Orchid Society Bulletin* 39:32–43, 207–220.

————. 1975. Stomatal development in *Ludisia discolor* (Orchidaceae): mesoperigenous subsidiary cells in the monocotyledons. *Taxon* 24:281–288.

————. 1979. Subsidiary cells in the Orchidaceae: their general distribution with special reference to development in the Oncidieae. *Botanical Journal of the Linnean Society* 78:41–66.

Williams, N. H., and C. R. Broome. 1976. Scanning electron microscope studies of orchid pollen. *American Orchid Society Bulletin* 45:699–707.

Williams, N. H., and C. H. Dodson. 1972. Selective attraction of male euglossine bees to orchid floral fragrances and its importance in long distance pollen flow. *Evolution* 26:84–95.

Williams, N. H., and R. L. Dressler. In prep. Generic considerations in the Oncidiinae (Orchidaceae).

Williamson, G. 1977. *The orchids of south Central Africa.* New York: David McKay Company.

Willis, J. C. 1973. *A dictionary of the flowering plants and ferns,* 8th ed. Cambridge: Cambridge University Press.

Wilson, E. O., F. M. Carpenter and W. L. Brown, Jr. 1967. The first Mesozoic ants, with the description of a new subfamily. *Psyche* 74:1–19.

Wilson, W. W. 1961. Selenocypripedium confusion. *American Orchid Society Bulletin* 30:806–807.

Winkler, H. 1906. Ueber den Blütendimorphismus von *Renanthera Lowii* Rchb. fil. *Annales du Jardin Botanique de Buitenzorg* 20:1–12.

Wirth, M. 1964. Supraspecific variation and classification in the Oncidiinae (Orchidaceae). Ph.D. diss. Washington University, St. Louis.

Wirth, M., G. F. Estabrook and D. J. Rogers. 1966. A graph theory model for systematic biology, with an example for the Oncidiinae (Orchidaceae). *Systematic Zoology* 15:59–69.

Yong, H.-S. 1976. Flower mantis. *Nature Malaysiana* 1(1):32–35.

Zeuner, F. E., and F. J. Manning. 1976. A monograph on fossil bees (Hymenoptera: Apoidea). *Bulletin of the British Museum of Natural History: Geology* 27:149–268.

Annotated List
of Orchid Floras

This appendix includes books devoted mainly to orchids, though in the tropical areas these may be sections or volumes of larger floras. I have listed primarily the works of the last few decades, which may still be used for the identification of orchids with a fair percentage of success. In most cases, mere lists have been excluded. All of the works are more fully cited in the bibliography.

United States and Canada

Correll, D. S. 1950. *Native orchids of North America north of Mexico.* [A useful flora, recently reprinted; well illustrated with drawings.]

Luer, C. A. 1972. *The native orchids of Florida.*

————. 1975. *The native orchids of the United States and Canada, excluding Florida.* [The two volumes of Luer give excellent coverage for this area. All species are illustrated by color photographs taken in nature.]

Mexico and Central America

Ames, O., and D. S. Correll. 1952–1953. Orchids of Guatemala. *Fieldiana: Botany* 26:1–727. [Still one of the best floras in this area. The keys work well, and there are a number of drawings.]

Hamer, F. 1974. *Las orquídeas de El Salvador.* 2 vols. [English, Spanish, and German. A good flora, with many drawings and a number of color photographs.]

Williams, L. O. 1951. The Orchidaceae of Mexico. *Ceiba* 2:1–321. [Without drawings or descriptions, and now rather incomplete.]

Williams, L. O. 1956. An enumeration of the Orchidaceae of Central America, British Honduras and Panama. *Ceiba* 5:1–256. [A list only.]

Williams, L. O., and P. A. Allen. 1946–1949. Orchidaceae. In *Flora of Panama.* [With descriptions and some drawings, but very incomplete. This flora includes only about half of the orchids now known from Panama.]

West Indies (except Trinidad)

Adams, C. D. 1972. Orchidaceae. In *Flowering plants of Jamaica.*

Fawcett, W., and A. B. Rendle. 1910. Orchidaceae. In *Flora of Jamaica.* [An old flora, but still one of the best, as the authors knew their orchids as living plants; with a number of drawings. Adams (above) brings the flora up to date.]

Garay, L. A., and H. R. Sweet. 1974. Orchidaceae. In *Flora of the Lesser Antilles.* [A good, recent flora with a number of drawings.]

León, Bro. 1946. Orquídeas. In *Flora de Cuba.* [Spanish. There are no good, modern orchid floras for Cuba, Hispaniola or Puerto Rico.]

South America

Dunsterville, G. C. K., and L. A. Garay. 1959–1976. *Venezuelan orchids illustrated.* 6 vols. [Not a flora, but gives excellent drawings of most Venezuelan species.]

Foldats, E. 1969–1970. Orchidaceae. In *Flora de Venezuela.* 5 pts. [Spanish. A very good flora, with many drawings and an excellent, illustrated key.]

Garay, L. A. 1978. Orchidaceae, pt. 1. In *Flora of Ecuador.* [This first part covers all orchids with soft, mealy pollinia; includes keys, descriptions, and some drawings.]

Ortiz Valdivieso, P. 1976. *Orquideas de Colombia.* [Spanish. This work has keys to genera, and illustrates each genus known from Colombia with a simple line drawing. There are also brief descriptions and some color photographs.]

Pabst, G. F. J., and F. Dungs. 1975–1977. *Orchidaceae Brasilienses.* 2 vols. [Portuguese, English, and German. Without descriptions, but with line drawings of all species (from dried specimens), and color plates of some.]

Schultes, R. E. 1960. *Native orchids of Trinidad and Tobago.* [A fairly complete flora with some drawings and good black and white photographs.]

Schweinfurth, C. 1958–1959. Orchids of Peru. *Fieldiana: Botany* 30:1–531. [Somewhat incomplete, and the keys are hard to use.]

Europe and Near East

Danesch, E., and O. Danesch. 1969–1972. *Orchideen Europas.* 3 vols. [German. Beautifully illustrated with color photographs.]

Duperrex, A. 1961. *Orchids of Europe.* [English, also in French. Well illustrated, a useful, field-guide-sized book.]

Landwehr, J. 1977. *Wilde Orchideeën von Europa.* 2 vols. [Dutch. A complete, recent flora for all Europe. Each species is illustrated with a color plate, and there are many drawings and vignettes. One hopes that it will be translated to other European languages.]

Renz, J. 1978. Orchidacea. In *Flora Iranica*. [English. A beautiful flora for a relatively orchid-poor area; with a number of very good color photographs.]

Summerhayes, V. S. 1951. *Wild orchids of Britain*. [A very complete flora with a number of good photographs.]

Sundermann, H. 1975. *Europäische und mediterrane Orchideen*. [German. A small, almost pocket-sized flora; very well illustrated with photographs.]

Africa

Ball, J. S. 1978. *Southern African epiphytic orchids*. [A large volume with maps and color plates for all species.]

Morris, B. 1970. *The epiphytic orchids of Malawi*. [With keys, descriptions, and some drawings.]

Perrier de la Bathie, H. 1939–1941. Orchidées. In *Flore de Madagascar*. [French. Keys, descriptions, and some drawings.]

Piers, F. 1968. *Orchids of East Africa*. [Not a flora, but useful.]

Schelpe, E. 1966. *An introduction to the South African orchids*. [Gives drawings and color photographs of genera.]

Stewart, J., and B. Campbell. 1970. *Orchids of tropical Africa*. [Not a complete flora, but has beautiful color photographs of representative orchids.]

Summerhayes, V. S. 1968. Orchidaceae, pt. 1. In *Flora of Tropical East Africa*. [Covers the Orchideae (including Diseae), with Keys, descriptions, and some drawings.]

Williamson, G. 1977. *The orchids of South Central Africa*. [Good, with many drawings and some good colored photographs.]

Northern Asia

Garay, L. A., and H. R. Sweet. 1974. *Orchids of Southern Ryukyu Islands*. [A recent and complete flora, with some drawings.]

Hu, S. Y. 1977. *The genera of Orchidaceae in Hong Kong*. [Gives keys to species, with a number of drawings.]

Lin, T. P. 1975–   . *Native orchids of Taiwan*. 2 vols. [Chinese and English. Not organized as a flora, but gives good drawings and many photographs; quite useful.]

Maekawa, F. 1971. *The wild orchids of Japan in colour*. [Japanese with English summary. A beautiful book, with excellent paintings, but hard to obtain.]

Tropical Asia

Holttum, R. E. 1957. *Orchids of Malaya*. [One of the most useful floras, with keys, descriptions, and some drawings.]

Millar, A. 1978. *Orchids of Papua New Guinea: an introduction*. [Just what the title indicates, but very useful, with many good color photographs.]

Pradhan, U. C. 1976. *Indian orchids: guide to identification and culture.* Vol. 1. [With keys, descriptions, and some drawings.]

Santapau, H., and Z. Kapadia. 1966. *The orchids of Bombay.* [The book is physically a bit flimsy, but otherwise well done, with keys, descriptions, and some drawings.]

Seidenfaden, G., and T. Smitinand. 1959–1965. *The orchids of Thailand: a preliminary list.* 6 pts. [A very useful flora, which does not pretend to be complete, with keys, drawings, and some colored plates. Seidenfaden has been revising parts of the flora as regional revisions.]

Australasia

Dockrill, A. W. 1969. *Australian indigenous orchids.* Vol. 1: *The epiphytes and tropical terrestrial species.* [A very good flora; all species are illustrated by good drawings.]

Hallé, N. 1977. Orchidacées. In *Flore de la Nouvelle Caledonie et dependances.* [French. One of the best modern orchid floras; each species is illustrated by an excellent line drawing and mapped, showing, too, the flowering times and altitudinal distribution.]

Nicholls, W. H. 1969. *Orchids of Australia.* [A beautiful book, each species illustrated by a colored plate.]

Glossary

In this glossary, I have tried to list those botanical and biological terms that have been used in the book, and all special terms which apply to orchids, whether I have used them or not. Those that are primarily orchid terms are indicated by (o). Some of the terms, such as acranthous, pleuranthous, and especially heteroblastic and homoblastic, seem to me to be unnecessary, but I have listed them since they are often encountered in orchid studies.

ABAXIAL   With reference to leaves and leaflike organs, the side away from the stem, normally the lower surface

ABSCISSION   The process by which leaves (or other organs) fall from a plant; a special layer (abscission layer) of easily broken cells is formed at the base, and the organ then falls off

ACRANTHOUS (o)   Having a terminal inflorescence

ACROTONIC (o)   Having the apex of the anther and pollinia associated with the rostellum or viscidium, as in the Spiranthoideae

ADAPTIVE RADIATION   The process by which a single group of organisms becomes adapted to a number of habitats, niches, or ways of life; generally said of any diverse group which has descended from a common ancestor

ADAXIAL   With reference to leaves and leaflike organs, the side toward the stem, normally the upper surface

AERIAL   In or having to do with air, as aerial roots

ALLELOPATHY   The production of inhibitory chemicals by a plant; these generally prevent or stunt the growth of neighboring plants

ANALOGOUS   Related in function or form, but not by origin (the caudicle and the stipe are analogous in function, but not homologous)

ANDROCLINIUM (o)   See CLINANDRIUM

ANGIOSPERM   Any member of the flowering plants; that is, plants with seeds, more or less closed ovaries, and flowers, class Angiospermae

ANGRAECOID (o)   Any plant resembling the genus *Angraecum*; any member of the subtribes Angraecinae and Aerangidinae

ANTENNA   In orchids, usually applied to the "trigger" in the *Catasetum* flower, an extension of the rostellum; when it is moved, the viscidium is forcibly discharged

ANTHER   That part of the flower which produces pollen

ANTHESIS   Flowering, the period when the flowers are open, or the opening of the flowers

ANTRORSE   Projecting forward, as in column wings

APOMIXIS   A process by which flowers produce seeds without any sexual union of cells; the embryo is derived from maternal tissues and is genetically identical to the mother plant

APPENDAGE   See CAUDICLE

ARTICULATE   Jointed; said of leaves or other parts which have an abscission layer or a joint at the base

AURICLE (o)   A small lateral outgrowth on the anther of the Orchideae

AUTOGAMY   Self-pollination without the aid of insects or other agents

AUTOPHYTE   Any green plant which manufactures its own food, as contrasted with saprophyte

AXIL   The angle between a leaf and a stem; a bud is normally formed there; axillary

AXILE   Pertaining to the axis; usually used to refer to the central location of placentation in an ovary

BASITONIC (o)   Having the base of the anther or pollinia associated with the rostellum or viscidium, as in the Orchideae

BLADE   The flat portion of any foliar organ, as in leaf or lip

BRACT   A scale- or sheathlike structure, homologous with leaf, but lacking the blade

BRYOPHYTE   A mosslike plant; a general term for mosses, liverworts, and hornworts

BURSICLE (o)   A sacklike covering over the viscidium in some Orchidinae

CALLUS (o)   A crest or fleshy outgrowth of the lip

CALYCULUS   A small cup or circle of bractlike structures outside of the sepals, as in *Epistephium, Lecanorchis*

CAMBIUM   In dicotyledon trees, the layer of cells between bark and wood, which produces new wood on the inside and new bark on the outside

CAPILLETIUM   In fungi, a network of sterile hairs or fibers in the sporangium; in orchids, the sterile hairs within the fruit, especially of epiphytes

CARPEL   A leaflike structure which bears ovules and seeds; in the orchids, the three carpels are so united and modified as to be nearly unrecognizable

CAUDICLE (o)   A slender, mealy, or elastic extension of the pollinium, or a mealy portion at one end of the pollinium; the structure is part of the pollen mass, and is produced within the anther

CHROMOSOMES   Rodlike bodies of definite number, which became visible during cell division and are made up of genetic material; an individual receives one set of chromosomes from each parent, and these are duplicated in every body cell; when the cells are not dividing, these bodies form a tangled network within the nucleus

CIRRHUS   See ANTENNA

CLAVATE   Club-shaped, like a very narrow teardrop, pointed at one end and rounded at the other

CLEISTOGAMOUS   Refers to flowers which automatically self-pollinate without opening; a form of autogamy

CLINANDRIUM (o)   The anther bed; that portion of the column under, or surrounding, the anther

CLONE   The aggregate of individuals derived by vegetative division or asexual reproduction from a single organism; all members of a clone are genetically identical

COHESION STRANDS (o)   Strands of sporopollenin that enclose or reinforce soft pollinia

COLUMN (o)   The central structure in an orchid flower, made up of the style and the filaments of 1 or more anthers

COLUMN FOOT (o)   A ventral extension of the base of the column which has the lip attached at its tip

COLUMN WING (o)   A wing- or armlike appendage of the column, usually lateral

COMPATIBLE   Said of any two individuals or species which may be crossed to produce viable seeds; an individual may be either self-compatible or self-incompatible, in which case it will not produce seed when pollinated with its own pollen (or pollen from the same clone)

CONDUPLICATE (o)   Of leaflike organs, with a single median fold, with each half being flat

CONGENERIC   Belonging to the same genus

CONNECTIVE   The sterile portion of the anther between the two anther cells

CONVERGENCE   When similar habitats or ways of life cause different organisms to resemble each other (not mimicry)

CONVOLUTE (o)   Refers to the way in which leaves are folded during development, rolled

CORDATE   Heart-shaped with the base at the broad, notched end

CORM   A thick, underground stem, usually of several internodes, as in *Gladiolus* and *Bletia*

CORTEX   In the root or stem, the tissues between stele and epidermis

CRETACEOUS   The last major division of the Mesozoic era, from 70 to 135 million years before present, during which the flowering plants evolved

CUNICULUS (o)   A nectary that is concealed within the ovary or pedicel

CYME   A determinate inflorescence, in which the first flower to develop is terminal; then bud(s) below that flower produce flower(s), and so on; this form of branching can continue to form a large and complex inflorescence

CYTOPLASM   The living material within a cell

DETERMINATE   A habit of growth in which each unit has a limited growth; See SYMPODIAL

DICOTYLEDON   Any member of the Angiosperm subclass Dicotyledonae, characterized by having net-veined leaves and two seed-leaves, as opposed to monocotyledon

DIPLOID   Having the normal two sets of chromosomes; sex cells have only one set (haploid); individuals with four sets are tetraploid, and so on

DISPERSAL   The movement of seeds or other reproductive structures from one place to another, by whatever agent

DISTAL   Away from the base, toward the apex

DISTICHOUS   Having leaves or other organs in two opposite rows, as opposed to spiral or whorled

DORSAL   Refers to the upper side of the flower; in orchids, technically the abaxial side of the flower, because of resupination

DROPPER   See ROOT-STEM TUBEROID

DUPLICATE   Refers to the folding of leaves during development; folded once, with each half flat

ELAIOPHORE   A gland producing oil

ELASTOVISCIN (O)   A clear, very elastic substance found in pollinia and especially in caudicles; this substance has not been chemically identified

ELFIN FOREST   A type of cloud forest in which the trees are dwarfed, usually with many epiphytes

ELLIPSOID   Spindle-shaped; narrow and tapering at the ends, three-dimensional

ELLIPTIC   Spindle-shaped, two-dimensional

EMBRYO   The young plant while still surrounded by maternal tissues, or within the seed

ENDOTHECIUM   In the anther, a layer of cells immediately beneath the epidermis, often with thickened cell walls

ENTIRE   With a smooth edge, as contrasted to toothed or lobed

EPHEMERAL   Of flowers, referring to those which last only a few hours, or less

EPICHILE (O)   The terminal portion of a complex lip, as in *Stanhopea*

EPIDENDROID (O)   Any orchid of the subfamily Epidendroideae

EPIPHYTE   Any plant which grows upon another plant

EPIPHYTOSIS   In trees, the "disease" caused by too many epiphytes, or epiphytes that are detrimental to the host

EQUITANT   Said of conduplicate or laterally flattened leaves which overlap each other in two ranks

EXINE   The outer wall of a pollen grain or spore

EXODERMIS   The outer cell layer of the root cortex

EXTRAFLORAL   Occurring outside of a flower; refers to glands that are found on stems, leaves, or outside of flower buds

FAA   The standard liquid preservative for plant materials: 60–70% ethyl alcohol with 5% glacial acetic acid and 5% commercial-strength formaldehyde; a small amount of glycerine may also be added

FIDELITY   In pollination, the tendency of a pollinator to selectively visit other flowers of the same kind, whatever the "motivation"

FILAMENT   The slender, sterile portion of the stamen which bears the anther; part of the column in most orchids

FLORA   The array of plant species in a region, or a book describing and enumerating the plants of a region

FUNICLE   The cord or thread which attaches an ovule or a seed to the placenta

GENUS (pl. GENERA)   A taxonomic category above the species; the generic name forms the first part of the species name

GLAND   See VISCIDIUM

GRANULAR   Refers to soft or mealy pollen

GREGARIOUS FLOWERING   Refers to orchids or other plants with short-lived flowers that correlate their flowering day by some environmental cue

GULLET FLOWER   In pollination ecology, any flower which forms a chamber into which the pollinator must enter

GYMNOSPERM   Any of the several types of seed plants which do not produce flowers, as conifers, cycads, and several fossil groups

GYNANDRIUM   See COLUMN

GYNOSTEGIUM   See COLUMN

GYNOSTEMIUM   See COLUMN

HEMIPOLLINARIUM (o)   In an orchid flower with two distinct viscidia, a single viscidium with the attached stipe or caudicle and pollinium (or pollinia)

HERBARIUM   A botanical museum; a collection of pressed and dried plant specimens

HERMAPHRODITIC   Having both sexes in the same flower; bisexual

HETEROBLASTIC (o).   Having pseudobulbs of a single internode

HOMOBLASTIC (o)   Having pseudobulbs of several internodes

HOMOLOGOUS   Corresponding in origin and in type of structure but not necessarily in function (bird's wings and human arms are homologous, bird's wings and insect wings are not)

HUMUS   Decomposing vegetation, leaf mould, or the soil formed by the decomposition of vegetation

HYALINE   Glasslike, translucent

HYGROSCOPIC   Capable of absorbing moisture from the atmosphere; in some cases, movements caused by such absorption

HYPHAE   Individual filaments of a fungal body; see MYCELIUM

HYPOCHILE (o)   The basal part of a complex lip, as in *Stanhopea*

INCUMBENT (o)   Refers to an anther that bends downward during the development of the flower

INFLORESCENCE   The flower (if solitary) or flower cluster of a plant

INTERNODE   The section of a stem between two nodes

ISTHMUS   A narrow portion of a lip or petal

KEYHOLE FLOWER   In pollination ecology, any flower in which the pollinator must place its tongue or proboscis through a narrow opening

LABELLUM   See LIP

LATERAL   To either side of a vertical line drawn in the center of a bilaterally symmetrical structure

LIGULE   A straplike body, especially an appendage

LIMB   See BLADE

LIP   One of the three petals which is usually larger and different in shape from the other two; the median petal

LITHOPHYTE   Any plant growing on rocks

LOCULE   A chamber in a closed structure, as in fruit or anther

MASSULA (pl. MASSULAE)   A packet of pollen in those genera in which the pollinium is subdivided into small packets; see SECTILE

MEDIAN   On the midline of a bilaterally symmetrical organ, as opposed to lateral

MENTUM (O)   A chinlike extension at the base of the flower, made up of the column foot and the lateral sepals

MERISTEMATIC   Tissue which retains the capacity for further growth; in monocots, normally in buds and at growing points

MERISTEMOID   In the developing leaf, the cell from which a stomate develops; guard cell mother cell

MESOCHILE   The middle portion of a complex lip, as in *Stanhopea*

MESOGENOUS   Having the subsidiary cell or cells all derived from the same cell as the guard cells

MESOPERIGENOUS   Having one or more subsidiary cells derived from the same cell as the guard cells, and the other or others from a different cell(s)

MESOTONIC (O)   Having the middle of the anther or pollinia associated with the rostellum or viscidium

METACENTRIC   Refers to chromosomes with the centromere more or less central; each chromosome with two arms

MIMICRY   Refers to resemblance between different kinds of organisms in which natural selection has favored the resemblance, as such; in the best known cases, an insect which is harmless or palatable may mimic a stinging, poisonous, or unpalatable insect; predators learn to avoid the common species (the model), and the mimic benefits from this learning

MONAD   A single pollen grain, as opposed to tetrad

MONANDROUS (O)   Having a single anther, said of all orchid subfamilies except Apostasioideae and Cypripedioideae

MONOCOTYLEDON   Any member of the angiosperm subclass Monocotyledonae, plants usually characterized by parallel venation and a single seed leaf (none in orchids)

MONOPHYLETIC   Descended from a common ancestor, as opposed to polyphyletic

MONOPODIAL   A growth habit in which the stem may continue to grow indefinitely, as opposed to sympodial

MUTUALISM   An ecological relationship between members of two different species in which both species obtain some benefit from the relationship; symbiosis

MYCELIUM   The plant body of a fungus, a tangle of threads

MYCORRHIZA   A symbiotic relationship between vascular plant roots and fungi

MYCOTROPHIC   Refers to vascular plants which obtain their food from organic material in the substrate through mycorrhizal fungi; saprophytic

MYRMECOPHYTE   Any plant which is associated with ants; it may grow on an ant nest, or the plant itself may provide a structure which is inhabited by the ants

NAKED (O)   Of pollinia, without any caudicles or other appendages

NECTARY   A nectar-producing structure or gland

NEOTTIOID (O)   Any terrestrial orchid with an erect anther and soft pollinia; any member of the subfamilies Spiranthoideae or Orchidoideae

NODE   The point on a stem at which a leaf or bract is attached

NODULAR   In reference to tuberoids, those which make up only a section of the root, as contrasted with those in which the whole root is thickened

OLIGOLECTIC   Of flower-visiting insects, visiting only a few (related) plant species (contrasts with monolectic and polylectic)

OPERCULATE   Lidlike

OSMOPHORE   A scent-producing gland

OVARY   The part of the flower which develops into the fruit

OVATE   Egg-shaped in outline, two-dimensional

OVOID   Egg-shaped, three-dimensional

OVULE   An embryonic seed

PANICLE   A branched inflorescence

PARALLELISM   The independent evolution of similar features in different organisms

PARENCHYMA   A tissue made up of similar, thin-walled cells, sometimes a storage tissue

PARIETAL   Borne on the wall, usually applied to the placement of ovules in an ovary

PEDICEL   The stem which supports an individual flower, usually jointed at the base, above the floral bract

PEDUNCLE   The stem that supports a solitary flower or an inflorescence

PELORIC   An abnormality in which the lip is like the other petals in form, or vice versa; a radially symmetrical mutant of a species which normally has bilaterally symmetrical flowers

PERIANTH   A collective term for sepals and petals, together

PERIGENOUS   Having the subsidiary cells all derived from different cells than the guard cells

PETAL   Commonly a white or colored (rarely green) flower part, borne within the sepals; in orchids, two of the three inner perianth parts, the third being called the lip

PETIOLE   The narrow, stemlike basal portion of a leaf

PHAGOCYTE   A cell which enfolds and "eats" solid particles, as a white blood cell

PHENOLOGY   The study of seasonality, as in flowering or fruiting

PHYLOGENY   History and development of a group through geological time; a scheme or diagram representing the phylogeny of a group

PHYLETIC   Pertaining to phylogeny or long-term evolution

PISTIL   The female, or seed-bearing, element in a flower; the ovary, style, and stigma

PISTILLATE   A female flower; usually said of fertile flowers that lack anthers

PLACENTA   That portion of the ovary that bears the ovules

PLEISTOCENE   The Ice Age, or glacial epoch, from 10 thousand to 1 million years before present

PLEURANTHOUS (O)   Bearing lateral inflorescences

PLICATE   Usually refers to leaves having several or many major longitudinal veins and usually folded at each one: pleated

POLLEN   One-celled spores borne in the anther; these develop the male gametes, or sperm nuclei

POLLINARIUM (O)   The complete set of pollinia from an anther, with associated parts, viscidium or viscidia and stipe

POLLINATE   The act of placing pollen or pollinia in a stigma

POLLINIUM (pl. POLLINIA (O))   A more or less compact and coherent mass of pollen; usually the contents of an anther cell, or of one half of an anther cell

POLYPHYLETIC   Refers to an artificial group, one that has descended from two or more different ancestral groups

POLYSTELIC   Having several vascular cylinders within a single structure, as the root-stem tuberoid

PRIMARY   Refers to the first formed root or stem; all branches of these are then secondary; in dicots, the plant body before thickening, the additional tissues being secondary

PRIMORDIUM   Any organ in its earliest stages of development

PROSCOLLA (O)   See VISCIDIUM

PROTOCORM (O)   A germinating orchid seed, the body from which a shoot and roots are formed

PSEUDOBULB (O)   A thickened stem, usually aerial

PSEUDOCOPULATION   A special type of mimicry, in which flowers resemble female insects and are pollinated by the males when these attempt to copulate with the flowers

PTERIDOPHYTE   A fernlike plant; a collective term for a level of evolution, ferns, horsetails, lycopodiums, and selaginellas

RACEME   An unbranched inflorescence in which the flowers normally open from the base upward

RACHIS   The axis of an inflorescence to which the pedicels are attached, above the peduncle

RAPHIDES   Needlelike crystals, usually of calcium oxalate; many orchid cells contain bundles of raphides

RECLINATE   Of anthers, bent away from the axis of the column, as in *Disa*

RECUMBENT   Bent back until the apex is below the base, as the anther of *Satyrium*

RESPIRATION   The process by which living cells oxidize food and produce energy, with water and carbon dioxide as byproducts

RESTRICTION   In pollination ecology, any devices which limit access to the nectar (or other reward) so that only the "legitimate" pollinator can obtain the reward

RESUPINATE (O)   The flower having the lip on the lower side

RETICULATE   Netlike, having veins which interconnect as in a net

RETINACULUM   See VISCIDIUM

RETRORSE   Recurving or pointing backward or toward the base, as in column wings

RHIZOME   A horizontal stem, usually on or in the substrate; in sympodial orchids, made up of the bases of successive shoots

ROOT-STEM TUBEROID   A storage organ (characteristic of the tribes Diseae, Diurideae, and Orchideae), primarily a swollen root, but with a bud and some stem structure at the base; this organ may push down into the soil to place the plant lower in the soil; the resting phase of the plant

ROSETTE   A densely clustered spiral of leaves, usually borne near the ground

ROSTELLUM (O)   A portion of the stigma which aids in gluing the pollinia to the pollinating agent; the tissue which separates the anther from the fertile stigma; sometimes beaklike

SACCATE   Sacklike, deeply concave

SAPROPHYTE   Any plant that does not manufacture its own food, but depends on organic matter in the soil, in vascular plants always in conjunction with fungi

SCAPE   In botany, the peduncle and the rachis of the inflorescence; in entomology, the basal segment of an insect's antenna

SECONDARY   Said of branches from the primary stem or root; in dicots, tissues produced by the cambium

SECTILE (O)   The condition in which soft, granular pollinia are subdivided into small packets, which are usually connected by elastic threads

SECUND   Having the flowers in an inflorescence twist so that they all face to one side

SEPAL   Usually the outer, green, perianth segments of a flower; in orchids, the outer three perianth segments, usually colored

SHEATH   A leaflike structure which enfolds a stem, pseudobulb, or young inflorescence

SHOOT   A growth, especially a new growth arising from the base of an older one

SIGMOID   S-shaped

SINKER   See ROOT-STEM TUBEROID

SPECIES   A population or group of populations which share a common pool of genetic and morphological features, which are interconnected by actual or potential gene exchange, and are separated from other such populations by barriers to gene exchange and by resultant gaps in morphological features; the basic unit in biological classification

SPECIATION   The process by which populations acquire sexual isolation barriers and become distinct species

SPIRANTHOID   Like *Spiranthes,* any member of the subfamily Spiranthoideae

SPOROPOLLENIN   The substance which makes up the exine of pollen grains; oxidative polymers of carotenoids and/or carotenoid esters; extremely resistant to decay or chemical change

SPUR   A slender, tubular or sacklike projection from a flower part, usually a nectary, commonly formed by the base of the lip

STAMEN   The male, or pollen-bearing, element in a flower, made up of filament and anther

STAMINATE   Bearing stamens, usually said of fertile flowers that lack a pistil

STAMINODE   A sterile stamen

STELE   The vascular tissues (support and plumbing) of a stem or root; normally forming an interconnecting cylinder within the organ

STELIDIUM   See COLUMN WING

STIGMA   The sticky, receptive part of the pistil, produces a viscid, sugary material which receives the pollinia and permits the pollen grains to germinate

STIPE (O)   A nonviscid band or strap of columnar tissue which connects the pollinia to the viscidium

STOMATE (pl. STOMATA)   The opening by which gas exchange occurs between the leaf and the atmosphere, each stomate is flanked by a pair of guard cells, which can open or close the stomate by changing shape

STYLE   The slender part of the pistil, which connects the ovary with the stigma; forms a part of the column in orchids

SUBSIDIARY CELL   A cell which is associated with a stomate and is structurally distinct from the other surrounding cells

SUBTEND   To extend under; subtending leaf, the leaf in whose axil a bud or inflorescence is formed

SUPERPOSED   One on top of the other; with respect to pollinia, flattened parallel with the clinandrium

SYMBIOSIS   See MUTUALISM

SYMPODIAL   A habit of growth in which each shoot has limited growth, new shoots usually arising from the base of older ones

SYNDROME   In pollination ecology, the complex of features which suggests adaptation to a particular class of pollinator

SYNONYM   A name which refers to the same group as another name, especially the younger (incorrect) synonym

SYNSEPAL (O)   A compound organ formed by the union of the two lateral sepals

TABULA INFRASTIGMATICA (O)   In *Oncidium* and related genera, a structure that is basal and ventral on the column, often convex and different in texture or color from the rest of the column; perhaps actually derived from the lip

TAXON (pl. TAXA) A taxonomic group of any rank

TAXONOMY The science of classification and naming

TELOCENTRIC Refers to a chromosome with terminal centromere; with only one arm

TERTIARY The geological period between the Cretaceous and the Pleistocene, from 1 million to 70 million years before present

TETRAD A unit of four pollen grains; pollen grains are produced in groups of four, and in some orchids they do not separate

THROAT The tubular portion of the lip, as in *Cattleya*

TROCHANTER A basal segment of an insect's leg

TROPHIC Pertaining to food

TUBER A thickened, underground stem, as in potato; not found in orchids

TUBEROID A thickened, tuberlike root

VANDOID (o) Like *Vanda,* any member of the subfamily Vandoideae

VASCULAR Pertaining to or with vessels or water-conducting tissue

VEGETATIVE Those parts and aspects of the plant not directly involved in flowering or fruiting; that is, roots, stems, and leaves

VELAMEN (o) One or more layers of spongy cells on the outside of a root; related to the epidermis in origin

VENTRAL On the lower side

VERNATION The way in which leaves are folded or rolled during development

VERSATILE In anthers, hinged and movable

VISCID DISK See VISCIDIUM

VISCIDIUM (pl. VISCIDIA) (o) A viscid part of the rostellum which is clearly defined and removed with the pollinia as a unit, and serves to attach the pollinia to an insect or other agent

VISCIN (o) See ELASTOVISCIN

VISCUS (o) The viscid glue produced by the rostellum, especially in flowers which do not have a viscidum; the term is rarely used

WHORL A circle of three or more leaves or leaflike organs attached at the same level on a stem

XERIC Pertaining to dryness or deserts

ZYGOMORPHIC Bilaterally symmetrical

Index